Arnaldo Arduini

Principles of Eidetics

Outline of a Theory

Springer-Verlag
Berlin Heidelberg New York
London Paris Tokyo
Hong Kong Barcelona
Budapest

Professor Dr. Arnaldo Arduini
Scuola Superiore di Studi Universitari
e di Perfezionamento S. Anna
via Carducci, 40
56100 Pisa, Italy

ISBN 978-3-642-49365-2 ISBN 978-3-642-49363-8 (eBook)
DOI 10.1007/978-3-642-49363-8

Library of Congress Cataloging-in-Publication Data
Arduini, Arnaldo.
Principles of eidetics: outline of a theory / Arnaldo Arduini. p. cm. Includes bibliographical references.

1. Thought and thinking - Physiological aspects. 2. Cognition - Physiological aspects. 3. Psychophysiology. 4. Neuropsychology. I. Title. BF441.A637 1992 153-dc20

Softcover reprint of the hardcover 1st edition 1992

Production Editor: Isolde Scherich
Typesetting: Camera ready by author
25/3140-5 4 3 2 1 0 - Printed on acid-free paper

To my wife Mariagrazia:

"I do love nothing in the world so well
"as you: is not that strange?
(Shakespeare, *Much ado about nothing*)

Preface

A number of themes in the previous book by this author, *Principles of Theoretical Neurophysiology*, are discussed further in the present one. The earlier work examined different physical points of view and outlined a comprehensive theory of nervous functions, starting from the basic assumption that the brain, an enormously complex organ, is a physical object and thus subject to the same physical laws that govern the "mode of being" of other, comparably complex systems. On this basis a general framework of laws, those of special and general relativity, was elaborated, promising not only metodological but also substantive value.

However, the brain is by all means more than a "machine" simply superintending the execution of the material functions, such as motion, sensing, and control of the internal organs of the body. The brain "thinks" and has "emotions," i.e., it performs functions that seem to escape physical definition. The aim of the present book is to provide a framework of principles for these activities. Though not physically graspable or measurable, these must still in some way be "implemented" in the brain structure in the form of neurons and their connections —in other words, they must possess a neurophysiology of their own.

This goal appears to recall the old mind-brain problem, but there has so far been no clear-cut definition of what one should be seeking in the brain when ideas and thought are concerned. In the maze of brain circuitry one finds only nerve impulses, measurable in terms of their number, time sequence, temporal organization, and the like; nevertheless, these forms of activity cannot be said to resemble an "idea" until one has obtained a notion of what an idea ought to look like as far as its "physical" structure is concerned. Thus, the problem has been tackled at the very root, i.e., what does one means by "idea," what is its "nature," what is its abstract "structure," how can this "abstract" structure be recognized within the brain's "physical" structure?

This search has carried the author farther afield than could be foreseen at the outstart. The title of this book by itself reveals the scope of the treated matter, which has expanded in the course of the endeavor to include many associated aspects of the problem. The discussion ends with a view of the eidetic function organized as a very comprehensive framework connecting conscience, consciousness, ideas, memory, thought, knowledge, logic, and language. However, the summary can propose nothing more than "principles," that is, merely the outlines of a theory and guidelines for future investigation —not yet sufficient for an exhaustive treatise, if indeed this is ever possible.

Contents

Introduction

Mephistopheles:
...vor allen andern Sachen,
Müßt Ihr Euch an die Metaphysik machen!
Da seht, daß Ihr tiefsinnig faßt,
Was in des Menschen Hirn nicht paßt;
(Goethe, *Faust,* v.1948-1951)

This spirit of evil is actually not that wicked, and substantively is not a liar; rather, he has his fun giving good advice accompanied by biting irony, alternating, as in the lines quoted from the interview with the Scholar. We shall try to be more clever than the young man, following the good advice but avoiding the pitfalls. Mephistopheles' words will remain as a "memento" throughout, as they should for anyone approaching the so-called higher functions of the brain. Indeed, anyone involved in investigating the brain becomes aware, sooner or later, of the missing link, that is, of the gap between detailed knowledge of nerve cell activity connected with somatic and autonomic functions, on the one hand, and vague presumptions, we should say, an almost total obscurity, about what concerns the mind and eidetic processes, on the other. We do not need to look far to find examples of unexplained phenomena: "will," for instance, as a mental event still lacks a physiological explanation, and has perhaps too many philosophical ones. However, there is no difficulty understanding conceptually and practically how the movement of a limb is produced in terms of nerve impulses generated in the motor areas of the cortex and relayed to the motor periphery. Much, probably, of the body functions, from the control of movement and the results of sensory events to the integration of somatic and autonomic activities, can be explained in terms of known synaptic organization representing a precisely ordered succession of neural events along preformed pathways. However, while we can follow to a great extent the flow of nerve impulses along the branches of sensory pathways and through the anatomical links connecting sensory and nonsensory centers, still we are completely at a loss when trying to describe sensation, its innermost nature, how it is actually aroused, and even whether is it representable at all in terms of nerve impulses.

One usually argues that sensations, emotions, will, learning and rationalizing can be understood better by the psychological approach, perhaps integrated with the monitoring of a number of neurophysiological parameters. However, psychology can deal only with behavioral responses to a stimulus or a "situation " What lies between these is generally the "black box" so appreciated by cyberneticians. While we may succeed in discovering the transfer function, this can only be in terms of numerical relationships between input and output; there is no evidence or even the lightest hint as to its nature or significance. Elsewhere, I have maintaned that the methods of neurophysiology are perfectly suited for an analytical approach, i.e., for discovering "mechanisms," but that these depend on general principles which cannot be investigated by analytical procedures. The only sensible approach to neuroscience must aim not at explaning brain functions in terms of nerve impulses, but rather the opposite; this is because the neural mechanisms are subserving the functions, and their organization must be considered as determined by the latter. Thus, the first goal should be to discover the principles upon which the higher (eidetic, intellectual) functions are founded and their relationships to the principles governing the somatic and autonomic functions. However, when rationality is involved, the problems of understanding brain functions not directly involved in somatic and autonomic "regulations" assume an unmatched complexity.

Increasing numbers of neuroscientists and philosophers are now approaching the so-called mind-brain problem, and such a substantial amount of literature has accumulated that still another monograph on this subject on the part of a neurophysiologist would scarcely seem needed. I would not have decided to write one were it not for my personal conviction that a great deal of misconception has come to obscure the nature of mind. On the one hand, there is still the vague belief in a mind-brain dualism (as most titles in this field indicate). On the other, the recently developed system of sophisticated "artificial intelligence" are being seen as models of the machinery operating in human intelligence. In most instances, mind has been taken as a self-contained entity without investigation of its essence or the wide gamut of all its properties. Rather, it has been approached only on the basis of some of its more obvious faculties —though probably those of a lower level— such as pattern recognition and categorization, which can also be reproduced in automata; learning and memory, on the other hand, have not been sufficiently investigated either in terms of their abstract importance or their substantive nature. (Regarding the mechanisms of the latter capacities of mind, as many interpretations may be found as there are books on the subject).

The eidetic problem, includes not only the formation of ideas, but also their use in reasoning, memory, building knowledge, and in will in general. This cannot be approached —at least not yet— with the methods currently used in experimental neurophysiology, and I would maintain that it is certainly not possible from the standpoint of a purely empirical experimenter. Before we can even conceive of an experimental approach, we must learn to ask the right questions, and this requires, in turn, a far deeper theoretical knowledge than is available at present. This subject requires a thoroughgoing treatment both from the point of view of brain capabilities and from that of the nature of ideas, as the fundament of all the mind's operations. Both neurophysiology and artificial intelligence models have so far tended to reduce mind's properties to the trivial capacity to recognize sensory patterns and to control overt behavior —as if these were significantly representative of rationality. Indeed, these capabilities may not even engage mind at all. An insect and an octopus are already far advanced in their ability to recognize sensory patterns, control behavior, memorize situations to be avoided, that is, in much of the field of so-called cognitive activities; regarding these lower-level functions, the ganglion-arranged nervous systems of these animals might find little to envy in the human brain. Everything in these lower systems is preprogrammed, as in a computer, and a computer equipped with artificial neuronal nets and suitable sensors can duplicate the behavior of the insect or the octopus in a perfectly mechanistic way —where the will and the freedom that are characteristic of rationality play no role. One may object that this is a matter of semantics, of defining the property "rationality." Yes, this is precisely our point of departure. But rather than give definitions in advance, we shall examine the essence of that which we call mind and its functional properties, i.e., the rationally based, so called "higher" nervous functions.

As in the case of the general brain function (in "*Principles of Theoretical Neurophysiology*", referred to hereafter as *PTN*), I maintain that the general framework of somatic and autonomic functions comprises the eidetic function as well, allowing and fostering a common interpretation of phenomena both of inanimate nature and within the living brain. This postulate may be tested, as it presents several differentiated approaches. In the "Appendix" of the same book, the common nature of some general features of these principles were discussed in the form of a confrontation between the today's neurophysiology and medieval philosophy. The present book outlines a theory of the higher nervous functions on the basis of the same few principles expounded previously. It will be clear that the missing link between the "physical" and the "eidetic" functions of the brain does not constitute a mere hypothesis to be submitted to confirmation but approaches the status of an apodictic, metaphysical postulate. Undemonstrable by physical means, this link is thus assumed as logically necessary. This may seem to be based on a contradiction. However, we must not forget that everything occurring in the brain —that is, involving the activity of neurons, the way in which we now understand this activity— is subject to the same physical principles as every other physical event. Our point is that when eidetics is concerned, there is no logical escape from the necessity of postulating, in addition, something of a different nature than nerve impulses. This is not to say that the eidetic processes of the brain do not rest on nervous activity as we know it; rather, it suggests the existence of other principles as well, not merely physiological in nature. These principles have not yet been described, and it is our endeavor to discover them and to reduce them to the level of the physical principles which we already grasp. Understanding the nature of these new, logical principles may require the recognition of hitherto unsuspected properties in the structural organization of the brain. In other words, the postulated "missing link" may not consist in presently unknown systems of connections and synaptic organization but in properties of known systems that we have so far failed to recognize. In these properties, I believe, lies the hidden connection between the abstractness of the mind and the materiality of the neurons.

From personal experience, I can foresee the neuroscientist's discomfort when asked to consider the unusual, the "immaterial," the "fleeting objects" that cannot be measured or observed in any rigorous way. However, when exploring mind, one cannot allow oneself to be hampered by the fear or the dislike of this type of unknown, of the regions where thought develops. Sooner or later the neuroscientist must face the problems of eidetic processes which he has so far left untouched, and which philosophers alone have explored; he must find them a place within an integrated, general framework of laws that also includes the noneidetic functions. The abundance of available literature on this subject might seem to deprive this observation of significance Many neuroscientists and philosophers are already engaged in the task, and from different viewpoints. The result, as Doktor Faust observed, was "und bin so klug als wie zuvor" —that is, I am no wiser than before. The fact is that after reading this literature, the missing link between neurophysiology and mind functions is still there, at least as far as I can discern. Gulfs continue to separate analyses of brain functions from historical-philosophical standpoints, psychological models of cognitive functions, neurophysiological models of pattern recognition mechanisms, and those of sensory-motor integration. Either one is attempting a philosophical systematization of mind, with little or no eye to the actual organization of the nervous structure, or, if one does do so, one is constantly hunting specific neural mechanisms for functions which may have little or nothing to do with true mind operations.

Churchland's 1986 book entitled *Neurophilosophy* presents a thorough, systematic review of the literature in this area, but unfortunately it was not available to me in time to aknowledge it in my own book (PTN). This was regrettable because I agree almost fully with Churchland's view. She affirms first the necessity of a multidisciplinary approach to so-called higher nervous functions: the domain of the neurophysiologist and that of the philosopher not separated by a sharp border but by a wide no-man'sland where knowledge and concepts from various disciplines are merging. At the present it is impossible to predict whether a physical-empirical, or a philosophical-theoretical approach will prove the more useful. Since according to Churchland, we cannot as yet say whether the relevant questions derive from the domain of physics and physiology or from that of philosophy, we must use both. This in turn requires that a single mind be engaged in the task by mastering two (at least initially) different heuristic procedures of reasoning, because it is the method of reasoning, not the amount of actual knowledge that is relevant. This, however, represents a second step; the first consists of clearly identifying the objects upon which we are focusing, and this actually represents the basis of the whole question. Before conceiving an experimental approach, we need to know the real object of the experiment; in other words, before we devote ourselves to practical measures, we must first consider the theoretical nature of the elements involved in the problems of mind.

Thus, there is a propaedeutic order that must be observed. Thematizing the first goal highlights the dilemma identified by Churchland. In examining the workings of mind, its means, the nature of its objects, we cannot say beforehand whether an empirical or a theoretical approach will be more fruitful, we must begin by selecting the best methods for making a rational choice. Let us consider, for example, a number of experimental facts that may be interpreted in various ways. Some neurons in the frontal cortex of the monkey are activated when an object in sight is located within arm's length. What does this mean? The neurons may, on the one hand, be considered simply as part of the sensory-motor integrating framework of operations, such as the agents activating the motor structures for grasping the objects in sight and that are reachable. On the other hand, we may regard the neurons as part of an eidetic process, for example, that involved in the perception: "Gee, look at that! I can easily grab it and play with it, and maybe eat it". The way in which we look at the phenomenon must be the result of a reasoned choice.

At this point my view diverges from that of Churchland. I disagree with her proposed solution of investigating the finer mechanisms of sensory-motor control (which in practice means investigating sensory-motor control with finer methods, in this case tensor analysis of neuronal activity) in the hope that this will supply hints as to the organization of mental states, as here the coevolutionary strategy would assume. I am thoroughly convinced, as Churchland is, that the laws and principles of physics are the same throughout matter, whether living or inanimate (my previous book was based on this view). While I am also convinced that in the case of somatic and autonomic functions tensor analysis is merely one of many possible methods of investigation, and that it can easily be substituted for by others, I strongly believe that only tensor analysis within a relativistic framework can be of any use in establishing the conditions of the development of mental states. In other words, it is quite reasonable to investigate every type of brain function by looking at the nerve events as forces acting in a continuum, and I also firmly believe that their processing *generally* follows the principles of tensor theory (a view shared for several years by Pellionisz and coll.). However, all this is far from accepting that the reality of the brain machinery may be reduced to the level of tensors empirically calculated from nerve impulse patterns in different structures. I use tensor terminology in the present book, and even more so than in the previous one, consider a special kind of tensor operator, the "metric tensor," to be responsible for the essential working principle of the brain, the reference framework transformation. Nevertheless, I do not forget that the brain has its own ways of making computations, ways that do not follow the heuristic steps of mathematicians; the neuronal circuits that the brain puts to work have little in common with those of computers, although in practical terms they may reach the same end. In short, the strategies are different. Moreover, the transitions between mental states are achieved by transformation of spatial *and* time coordinates that do not involve all the concerned parameters equally, which means that the brain continuum is a non-Euclidean, multidimensional space (and this calls for metric tensor.)

For this reason, I refrained in my previous book from discussing models of nervous structure organization that strictly mimick the circuitry of their physically assembled counterparts. (It might seem, however, that I actually have proposed a model by advancing the hypothesis of implicit implementation of the ideas; I intended only to advance the hypothesis of similarity of principles, leaving completely undecided the question of the actual circuitry and operations). Models, however, might be quite cleverly conceived for constituting it, for instance, the "phase space sandwiches" described at length by Churchland. For want of better definition of the nature of mind in general and of its workings, first of all its eidetic workings, I cannot accept the simple parallel between the organization of somatic functions (as motor control) and that shapeless phantom which mind still is for all of us. I take a more radical position and dispute the attribute "mental" regarding many so-called cognitive processes until a clear-cut, nonequivocal definition of them is given.

Since there seems to be no need for better definition of the substance of motor control, we should devote ourselves first to theoretically investigating mind.

The neuroscientist should turn philosopher? Yes, surely he should, provided Mephistopheles' words quoted at the beginning remain in his ear. This means that the philosophy concerned must be of very special kind, one that in order to avoid the pitfalls of fanciful reasoning is based on self-exploration of consciousness, and remains tightly anchored to the known processes of the actual living brain. (This would be the type of philosophy more congenial to Goethe, so different from Enlightenment philosophy, of Kant, from the "arid" philosophy of pure reason —the sort of philosophy that Hegel perhaps discussed with Goethe during the years of teaching in Jena, at the time when the *Phenomenology of Mind* was developed.) The relativistic framework seems best suited as it allows a more complete description of the physical principles of brain functions; similarly, when mind is concerned, the phenomenologically inspired investigation, in Hegel's sense, seems the best suited for arriving at a definition of the objects constituting it while also accomodating the relativistic description. The present book is not a revisitation of brain in terms of phenomenological philosophy; the reader will find references to the latter, from Hegel to Husserl, but not a systematic discussion of phenomenological tenets. Our attention is always focused on the brain, and the brain alone. A further and more practical reason for keeping an eye trained on philosophy is that the life sciences are applied sciences; they do not enjoy an independent theoretical basis of their own but must find their roots ultimately in physics. A systematic approach along these lines is presented in my previous monograph dealing with theoretical neurophysiology.

Here, however, mind is the topic, and the language of physics alone is not powerful enough. The truth is that we are not simply looking for "facts" but for the "mental instruments" that may enable us to approach our science, those instruments which have enabled scientists of the past first to outline and then develop the theories of physics and mathematics. These instruments constitute the language of philosophy, as we shall see in the lst chapters of the present monograph, that which we may consider the "internal language" of mind. In his recent book, *The Remembered Present*, Edelman deals with this specific problem. Regarding the idea of "concept," Hegel observed: "the concept alone...has actuality, and gives this actuality to itself" (in the very first paragraph of "Hegel's Philosophy of Right"). Edelman places this insight at the basis, or, better, at the origin of any study of eidetics. Following different paths of thought I have reached the same conclusion in the present book, and finding Hegel and Edelman to have preceded me in this was a great relief, since I consider myself anything but a true philosopher.

The reader desiring a review of the current issues in mind-brain problems is referred to Edelman. These issues are still very fluid, and the field is perhaps still cluttered with too many experimental "facts." As will become apparent in the following chapters, no comprehensive overview can be reached merely by assembling facts. Again, in the words of Hegel: "Ferner ist bei einem solchen Aggregate von Kenntnissen, das den Namen Wissenschaft nicht mit Recht führt, eine Konversation über Zweck und dergleichen Allgemeinheiten nicht von der historischen und begrifflosen Weise verschieden, worin von dem Inhalte selbst, diesen Nerven, Muskeln und so fort, gesprochen wird." (*Phänomenologie des Geistes,* "Vorrede"). This sentence clearly states that aggregates of facts or knowledges cannot by themselves make science if are not comprised in a conceptual framework. The same idea permeates the argument of Edelman's book, since what he strives to produce is a "system," not mere collection of facts: Hegel says (*ibidem*), that knowledge is effective, and can be represented, only as a system or as science. The psychological and the recently developed "neural net" models promise experimental support, however still within the limits of the cognitive aspects. Regarding the unlimited use of models and the faith in experimental tests I do not fully agree. On the one hand, the oversimplification of the problem implicit in models requires the acceptance of not clearly defined assumptions; on the other hand, the facts that an experimental test may offer can at the best be accepted as "circumstantial evidence," not as formal proof. Despite, or perhaps because of my long experience as an experimental neurophysiologist, I lend more credit to the results of reasoning than to those of observation.

Particularly in this area, no strictly experimental approach seems justified at the present without being rooted in a framework of principles encompassing the manifold aspects that the mind problem comprises, and no investigation can be properly designed if the necessary agreement with such principles is not present. This is what I endeavor to achieve in the following pages. Reaching the goal of a comprehensive view, even if incomplete in the details, was of course already in mind with the publication of the previous book. The present monograph is thus a natural continuation.

Rationale

Mephistopheles:
Gebraucht der Zeit, sie geht so schnell von hinnen,
Doch Ordnung lehrt Euch Zeit gewinnen.
Mein teurer Freund, ich rat'Euch drum
Zuerst Collegium Logicum.
(Goethe, *Faust*, v.1908-1911)

Hypothetical systems are often tempting in their explanation of the organization of functions. Beautifully packed and consistent in themselves, these systems lack a likely correspondence in terms of the operations which the nervous system may reasonably be assumed as capable of performing. If the hypothesized processes and the proposed solutions cannot find in the brain matter a likely basis for their realization, they must be discarded. The fact that a given model of functional organization looks appealing and facilitates explaining the phenomena does not make that model any more true and real unless the theoretical "necessity" of it alone can be demonstrated. The approach taken here will be as naive as possible. I do not analyze and discuss present theories, since there presently is no systematic theory available. I attempt rather to build one myself, seeking to penetrate into the essence of the phenomena treated; as we shall see, this is the only way in which a theory may be constructed. The major difficulty here lies in making sure that one is posing the right questions, these can stem only from knowledge of the essence of the problem's terms. This means that asking offhand questions about the mind-brain relationship is not a suitable starting point for confronting such an enormous problem. There is insufficient knowledge about the essence of brain functioning, on the one hand, and too many contrasting views about the essence of "mind," on the other. However, there is no harm in continuing to speak of mind provided we can assume that we are talking about the same thing, as it is intuitively understood.

The basis of my view of mind is the "real" existence of "abstract" entities. While at the moment this is no more than a postulate for which obviously no one can produce any practical demonstration, it will be apparent in the discussion that the real existence of abstract entities is a logical necessity, not as mere products of mind, i.e., as artificial constructs useful for indicating the results of brain cell operations, but as real, fundamental constituents of mind itself, those which actually make up the "substance" of mind. Nothing characterizes Man more faithfully than his mind. The dualism that still plagues neurosciences —the mind's two natures, spiritual and material— has been haunting the debates between theologians and scientists for centuries. The increasingly reinforced empirical-positivistic standpoint in modern science confused the issues by confounding the mechanisms of mind's operations, which are rendered visible in the form of cell activity, with the essential nature of mind itself. Unfortunately, though not surprisingly, today's powerful means of investigation, while enabling us to penetrate into the more subtle operations of nerve cells, are also surreptitiously leading to the conviction that the sole reality is that subject to our senses and detected by our measuring instruments. This occurs (obviously) because for want of a theoretical reference framework nothing else actually reaches the eyes of the mind. While recognizing that the sole physical reality might be explained —and only up to a point— as far as the physical brain is concerned, certainly it leads one to ignore any other reality that the physical means is unable to unveil, as well as any other type of reality the existence of which is denied according to the view above.

However, even if by hypothesis one concedes that the ideas, the thought, all the elements of eidetic functions and their properties are mere physical products of brain, no physical measuring device that we know of, or that we can conceive at the moment, could detect them as such, i.e., for what they are in se and per se. The sole reality that our instruments can detect is that of physicochemical events of the nerve cells and fibers, connected with the essence of the ideas and with thought in still completely unknown ways. The ideas are abstractions, as everyone, neuroscientist, philosopher, and theologian, would agree, and if they are seen as constituting the basis of mind, mind is also abstract. At the same time, the fact that only very few, or none, of the many and diverse physical activities of man escape mind's control shows that mind is deeply entrenched within the brain matter. However, if so, the relationship between mind and brain are in the reverse order, that is, the overall brain activity is made congruent with mind's abstract properties, and not only the activity of brain but also its structure is dependent on mind.

Thus, that which we call mind is first of all an abstract but otherwise hardly definable entity. One could spend as many words as one thinks it necessary and still not arrive at a precise, clearly understandable, and universally acceptable definition. There are too many the facets of mind and too many points of view from which to see it. Since it is an abstract entity, one cannot even "locate" it. The various properties attributed to mind that we call higher functions are essentially also abstract entities, such as knowledge, thought, memory, and language, although the close relationships of these with the physical brain cannot be questioned. Despite these links one cannot produce a rigorously logical demonstration that mind is confined to limited areas of brain; what is demonstrable is only that neural matter is indispensable for these properties to be expressed at all (in the sense that they may be experienced and used). Whether one regards brain as mind's instrument, or brain as mind itself in principle makes a great difference, but we shall not consider the question now. For the purposes of this monograph we can spare the philosophical controversy by assuming that in either case the problem remains that of how the functions of mind are performed in the brain, by the brain. In developing a theoretical approach to the brain function in general (in *PTN*), I maintained that any brain theory in order to be acceptable should be sufficiently general to account for the whole range of brain phenomena, including those pertaining to the intellectual domain. (On the other hand, on account of the close interdependence of all brain functions, as stressed by modern neuroscience, a very general validity is a prerequisite for any brain theory.)

Since mind remains a completely unknown abstract entity —I prefer to say abstract *structure*, for reasons presented below— the philosophical beliefs, or the simple points of view, of the writer cannot fail to influence the general way in which the matter is treated. We postpone until the end the more complete framework into which a theory of mind and of similar structures may be integrated. The description of mind as a unitary whole together with its properties under the term abstract structures, has the advantage of conveying the concept of immaterial entities having immaterial structures of their own. However, these must also be endowed with the property of being somehow embodied in brain matter, since only through the brain can they operate and thus also be subject to investigation. This is the only viewpoint that allows us to suspend the debate about brain as mind and brain as mind's tool, but it requires of the scientist, to start with, that he accept the real existence of such entities as the abstract structures, which is to be "demonstrated" below. Here begins the metaphysics of brain. I believe the necessary link between abstractness and materiality is represented by the concept of structure —a unity of elements, some alike but most different, each with its own specific purpose in relation to the purpose of the whole.

I propose the hypothesis that only by possessing a structure can the abstract entities be embodied in corresponding structures of physical nature, by a kind of topological transposition; we shall leave open for the time being whether this is isomorphic or not. At the basis of every further speculation on the subject is the question of the extent to which limitations inherent in the physical, structural properties of the brain influence our way of constructing abstract structures themselves since we must embody them in circuits of neurons. It seems obvious that the properties of any system, and therefore its capability for constructing some type of abstract structures, are limited by the possibilities of its present organization. Against this is the experience that the brain, through plasticity and adaptation, has the capability for changing to some extent its physical organization in response to changed conditions or demands. The demands imposed may be considered the most common factor determining every physical structure and therefore the possibility of brain in turn to embody the abstract structures. Rephrasing the question, we may ask what are the limits of our capability for building abstract structures, and where and how these limits are set. The question is, then, whether the physical structure of brain circuits addresses the fundamental intellectual functions —formation of ideas, thought, language— along predetermined paths. Rephrasing again, we may ask whether our intellect with its many abstract structures is fully dependent on (philogenetically or ontogenetically) predetermined physical structures or is free to move in a "space" unattainable by physical conditioning. The question of the freedom of mind, a more general property than mind's free will, so far has afforded only philosophical approaches. However, we need physiologically founded arguments to integrate freedom and free will into the complex of eidetic relationships between conscience, on the one hand, and consciousness-awareness, on the other. The link unifying philosophy and physiology lies in the Self, the *Ich*, the Ego, not as psychological abstractions, not as the ensemble of relationships of the individual with the world but as the substance itself of the individual, the *Ich-esse* to which the individual's essence refers.

Hefty problems, but, I fear, meager answers. As I said above, the starting point will be as naive as possible, in the sense that no predetermined philosophical framework is assumed; thus, we shall pursue our study without prejudices, solely on the basis of the present knowledge of the nervous system, with all its inherent limitations. Needless to say, however, a number of our arguments and conclusions will be found to fit more or less into some philosophical framework, but this is not the purpose, and it will not be stressed. Occasional references are made to well-known philosophical theories. For instance, although not being straightforwardly phenomenological, the approach here makes use of Husserl's methods, and some of the opinions expounded here echo Hegel's philosophy of right and the phenomenology of spirit. This convergence seems the more remarkable since a phenomenological approach to the study of brain and the approaches of present day neurophysiology start from different, if not opposed, views. The former has its center of gravity in the "self," the conscience and its continuous stream of experiences (*Erlebnisse*), outside any reference to the material substrate that makes all this possible. Our point of departure, rather, is the brain; thus, the properties pertaining to the self —such as its experiences, the states of consciousness, conscience— are always referred to the physical structure, not because they are by nature exclusively physical manifestations, but because the physical structure is necessary for us to experience them and to acquire knowledge of them. On the basis of the presumed topological transposition, the embodiment of the abstract entities in the brain's structure can be assumed in turn to give us some insight into their nonphysical structures.

However, besides being naive, at the beginning we shall assume a radical standpoint. Looking for a physical explanation, or for the physical bases of abstract structures, no distinction is made beforehand between the self with all its different aspects, and its abstract products such as the ideas; since it cannot be taken as demonstrated that they belong to different classes of objects, there seems no logically justified need for a distinction. Furthermore, the generally accepted unity of physical structure and psyche (with all its properties) does not constitute a psychological phenomenon but is a fact of the human brain life, thus belonging to the physical world, and affords no interpretations but only the acceptance of it as a reality. By treating the brain and related abstract structures, as we do throughout, as separate objects endowed with properties corresponding to unique, common functions, we merely analyze separately two aspects of the same object, aiming at understanding how they are interrelated, and how a unity emerges from them. With this, the material and the immaterial are indissolubly bound and, what is more, subject to common limitations. If we cannot distinguish between *cogitans* and *cogitatum* (the thinking self and its thought), the proposition *cogito ergo sum* can be expanded into *cogito, ergo sum cogitans simulac cogitatum*, which means that the "essence" of man, i.e., that which makes him what he is, is further determined for each individual by his thoughts. Such a perspective should not appear too restrictive for either material or immaterial properties. The matter of the body is finite and therefore limited in its abilities. If in practice the abstract structures cannot be separated from matter, the same qualities, finiteness and limitation, extend to them as well. On the other hand, the property of immateriality does not suffer such restrictions, and therefore, if our previous assumption is accepted, it also should not suffer the restriction of material properties. Against all appearances, there is no contradiction in these propositions. The properties exhibited by inanimate matter in general depend on a physical, submicroscopic organization, on the ensemble of laws holding together its infinitely small components (its charged and noncharged particles) and on physical parameters such as temperature, pressure, etc. The same is true in the case of the living nervous system. Its fundamental constituents, the neurons, are grouped in subsystems endowed with different properties by virtue of matter's macroscopic organization. However, in the nervous system the same nonmaterial properties expressed by that organization can foster other, new, and more advanced patterns of organization of matter, thus causing, in turn, the appearance of more sophisticated abstractions, provided the suitable stimuli, conditions, or situations are present. (In principle, there should be no need for assuming further growth of neural matter, but present neuroscience seems to hint at this possibility.)

Organization is part of the concept of structure, signifying a "logically" interconnected ensemble of elements; thus, organization is the source of logical structures from "raw" material. As in *PTN*, here too we are inclined to accept the conclusion that the functional organization is the factor that fosters the organization of matter, and therefore determines the latter's properties, for example, in defining the physical structure of the organs. This means that we need to arrive at an understanding of the principles of the organization and the origins of its motion. Certainly, putting all the burden onto organization is not a way of simplifying things. We shall see that the main questions are those concerning the logicality of the links that bind the elements of the eidetic constructions, on the one hand, and language, on the other: not two independent but rather two integrated aspects of a single question. It is no wonder that language has long since become the object of extended research by philosophers and psychologists, who regard it as the (apparently) more accessible of the higher functional properties of the human brain, the one hopefully representing the immediate, natural expression of an hidden plan of organization deeply rooted in philogenetic and ontogenetic development. But, here, distinction must be made between two "languages." On the one hand, language in the more common sense, that is, the verbal symbols with which we construct our thoughts, and to which we then give physical expressions. On the other hand, there is what we may call the innermost source of language, i.e., the abstract constructions of logic upon which language is founded and giving the latter a universal understandability. These two languages should not be confused with one another.

At the basis of the complex question concerning structure and organization is the logicality of the set of connections between the elements of the organized unity. Therefore, we should deal with the possible basis of logic itself, its definition in the context of our problem, its very essence in the procedures of embodying the abstract structures in the brain, what we might call its "mechanisms." The task is well beyond our reach. Little seems to be gained by assuming that logic is founded on a priori postulates, belonging exclusively to a region outside our mind and independent of brain (however this assumption might be thought to be hidden in Part V of this monograph). However, even if this is true, we can never demonstrate that the a priori postulates actually constitute an absolute reference (if the latter ever exists), because in any case we can gain consciousness only of what our intellect is able to conceive and thus to understand. We should not miss the point that thought, regardless of whether concerned with science or not, inasmuch as it is understandable is a product of our intellect and of the brain expressing it; that all the ideas and reasonings must be somehow materialized in the brain matter and are therefore conditioned to some unknown extent by the structural organization of the brain. It can only be speculated whether we shall eventually be able to ascertain the nature of this physical conditioning. Accordingly, our physical brain acts as a "filter" for whatever abstract structure is not conceived within itself, and does not let through that which does not fit its own logic for building and embodying abstract structures.

Recognizing the logicality of its own constructions (abstract and physical structures) would seem to represent the main endeavor of mind: an operation that can be conducted at different levels in different tasks, such as the trivial case of finding the logical place of an object in a physical ensemble, or of a number in a series, up to the links in sentences, and the conclusions of a reasoning chain. Although different and arbitrarily selected criteria of judgement may be used in different instances, there must be a basic, original requirement for all —since recognizing that logicality is a peculiar capacity of humans— so that mind may confidently found its judgements on a belief in its being logical. It is apparent that this judgement must have a double origin: the first genetically determined (we can safely assume this although we are not yet able to define it precisely), the other dependent on the acquired capability of mind (a function of training, experience, knowledge, etc.). A great deal of subjectivity is thus involved, but if we were able to neutralize, as it were, the variable and controllable acquired components, we should be able to recognize the basic structural principles of the logic of the human brain, those originally built on its organized structure.

Regarding the possibility of their being fully understood, we should not lose sight of the limitations inherent in the structures made up of a material and an abstract part, such as those of the "eidetic" brain. There is something in the eidetic processes, from idea formation to reasoning, that seems to escape our firm grasp. This has to do with the fact that we may consciously think only if thought is materialized in the form of brain cell activity (at least, this is the commonly accepted assumption). We shall further accept the concept that the idea is itself consciousness and always presupposes it; in this way, the self, the mind, and its so-called products also become real by being experienced as *Erlebnisse* of consciousness, and we can no longer distinguish in practice, as noted above, between self and thought, since at the level of the brain's consciousness they are not differentiable. However, this is not the whole story. There also are objects (in the general sense of objects of thought) that exist in some way outside (or below) the level of awareness (or full consciousness). For instance, while being aware of one object, we are not completely oblivious to the existence of other objects surrounding or related to the specific object of our thought, as well as objects of which we have no direct experience. It is not that we simply know theoretically that there are other objects besides those directly experienced. Those objects, inasmuch as we know their existence, are somehow present within us, in processes of our mind, and their existence cannot be denied; or, perhaps better, they are in a way lingering at an outer "fringe" of mind, although not yet crystallized in an expressed thought and thus not yet completely "real" objects of consciousness.

The latter question is not one of minor importance. One may perceive in it the hidden kernel of the unsolved problems of the mind-brain relationship. Accepting the existence of these fringe objects of mind, we must also pose the question as to the "space" to which they belong. If not the immediate and continuously renewed present of our exploring mind, should we posit for them the existence of another region? A region where past and future objects of consciousness —and thus of thought— in the form of abstract structures, though not presently realized completely in cell activities, are nonetheless present and endowed with some kind of reality of their own? If we accept that thought is becoming life and our brain mind life, or our life (the *cogito ergo sum*) as being coincident with our self, of what life do the objects outside thought, but not altogether outside consciousness, partake? If we maintain that in order to be perceived —even as a shadowy perception pertaining to a fringe— the abstract objects must somehow have a share in brain cell activity, the innumerable objects at the "horizon," and beyond, of our consciousness must somehow be part of our brain, although not quite in the same way as those directly experienced, perhaps relying on some as yet unknown properties of brain matter organization.

To avoid confusion, we must tentatively provide different names for the different situations: "awareness," the state of being directly and fully conscious of an object of our thought, comprising the capability for exploring it; and "consciousness," a poorly defined something, comprising all the existing objects, but still as shadows, not directly examined under the focus of mind, including the fringe, the horizon, and beyond. Thus, consciousness is not a state comparable to awareness, but indicates the simultaneous presence in mind of all those objects not directly explored by our conscious thinking. This problem relates to another, still more fundamental, question: the existence of abstract objects that cannot be directly experienced and dissected under the focus of consciousness (i.e., in awareness). This concerns the relationship between the concept and the idea, the epistemologically ordered sequence of the entities from which thought arises. On the one hand, the precedence of the concept, an abstract object offering itself to immediate —though incomplete— understanding, and self-explanatory, not needing proof of its truth inasmuch as it represents a primary reference for other objects dependent on it; thus it does not itself belong to the realm of causality but is a structure from which many others originate. On the other hand, the symbolized form of the concept, the idea or the ensemble of ideas originated from the latter and giving it, though incompletely, a definite "shape," a form embodied in the brain matter, a description as faithful as possible to its essence. Only the ideas can be submitted to the perusing inquiry of awareness because, embodied in the brain matter, they are objects of space time, of the causality domain, the sole domain in which awareness may be present.

It is clear that the two terms, consciousness and awareness, indicate two very different phenomena. The latter is a time-dependent phenomenon such as every process (or state, since a state is actually a process, see *PTN*) in which the activity of cells is involved, resulting in conscious thought. The former is a less time-bound phenomenon and represents the continuous presence of simultaneous knowledge (should properly be termed "conscience") that may or may not fall at some times under the focus of awareness. These are not new propositions but are contained already in the previous proposition on the existence of fringe objects, a halo of ghosts of objects outside direct awareness. The distinction between conscience, on the one hand, and consciousness, on the other, corresponds to a difference of "spaces." Consciousness may be thought of as constituting a sort of material "phase space" (I am actually thinking of a structure much more general throughout the brain than Churchland's "phase space sandwiches"), having all the properties of a mathematical phase-space but made up of all the concept-ideas that brain-bound mind has in itself. On the other hand, conscience is something that might also take the place of memory, although in a space that is altogether abstract because it is in no way bound to brain matter (which belongs to physical space time). Consciousness is no less complex a structure: it is idea and state-space at the same time. As state-space it must have a physical definition (still to be clearly described) in brain matter, although necessarily with a peculiar relationship to time since as a state it allows simultaneity of presence of the abstract objects in brain space time. (The problems connected with conscience, consciousness, and awareness is discussed extensively in Part V.)

These considerations again involve the questions of constancy and variability of brain functions, as well as the role of time in our central nervous system (discussed at length in *PTN*). Here, we have two clear instances of the different implications of the term constancy: on the one hand, a perhaps truly constant situation (as we commonly assume to be the existence of knowledge); on the other, the possibility of constancy of the outcome of time-dependent nervous processes, the possible constancy of thought given the same premises. If we maintain that everything connected with consciousness —even if in a very loose way, as we said of the objects forming its horizon— must also be in some obscure way connected with nerve cell activity, then a constancy of consciousness is obtained only by a steady-state situation of those cell activities. In consciousness, the objects are required only to exist, to be present, but since we have no way of exploring the content of consciousness without "disturbing" or altogether destroying the state —only awareness can explore, though one object at a time— we cannot be certain that the content of consciousness actually is maintained as constant throughout; as was noted above, it is a presumption but, we shall see, a necessary one.

Now, it seems that most arguments in this book may be conducted on the same bases as those in the previous one (*PTN*), i.e., the same arguments may be used also for problems not belonging to the somatic or autonomic sphere. What we have just said is that the abstract structures (ideas, thought, knowledge) in order to be "handled" in our intellectual operations also need to be somehow embodied in nerve cell circuit activity; no other way seems to exist. However, in this case there is something more, and fundamental. When constancy and repeatability of some aspects of the higher functions are required, we must face the question of phase space and memory, in such a way, however, that the latter does not become separated from the general problems of the neurology of eidetic processes but may become part of a single reference framework. The composite nature of eidetic structures, consisting of abstract though also matter-bound structures, leads us to the conclusion of the nonexistence of memory under the generally assumed form of "matter-imprinted," but rather for memory as being of the same "fleeting" nature as the ideas and linked to matter only in the same way as ideas are. Although the arguments we shall advance in favor of this interpretation may also appear strict, the resulting picture is still vague enough to allow further elaboration.

We must expect eidetic and the other related functions to be influenced by the same factors that affect the performance of somatic and autonomic functions, however more profoundly. These factors, in addition to emotionality, are represented mainly by what we have called broadly philosophical principles, abstract structures much more complex than simple ideas but not concepts in themselves, part of the eidetic wealth embodied in our brain and capable of bringing about state transformation of the latter. We find no boundaries for the mental processes when speculating jointly about philosophy and psychology. Being peculiar to man, psychic features cannot be investigated in animals, except to a very limited extent and for restricted purposes. The main problems of psychic life are in the domain of "spirituality," involving metaphysics in a broad sense. It is quite natural that so far only psychologists and philosophers have produced an abundant literature, but we shall not discuss their theories, nor follow them, although we are concerned with essentially the same problems. The reason for this decision is that psychological and philosophical theories, alone, lack the ability to take into proper account the real capacities of the human brain resulting from structural and functional organization. On the other hand, while a substantial, thorough philosophical knowledge seems a prerequisite for approaching the questions from a point of view that is not strictly neurophysiological, we shall try to do without this. Instead, we will account principally for neurophysiological properties and capabilities of brain matter, and in this way we avoid becoming involved in a critical discussion of theoretical sciences up to now. Thus, neurophysiology will save us from voluntarily and consciously adopting any one of the psychological, philosophical, ethical, or religious theories of today while at the same time leading naturally to the sought-for integration of the materially directed with the spiritually directed activities of the brain, which is our ultimate goal.

Such a complex matter as eidetics requires a much more extensive treatment than can be here provided. Although the time is certainly ripe for a fundamental consideration of eidetic problems, the presence of still unsurmountable difficulties in understanding the latter means that we cannot take them lightly, but that a more gradual approach must be preferred. After all, the declared purpose is to give an initial, possibly comprehensive view of the eidetic function, highlighting presumed links of its abstract properties with the matter-bound ones, still remaining within the framework proposed in *PTN* for brain function in general. The fulfillment of this first stage of our purpose requires starting by using basic terms such as thought, idea, concept, memory, knowledge, consciousness, and language in their common, quasicolloquial senses, letting their more precise connotations emerge gradually, as by themselves. In the course of the discussion, the terms are, so to speak, allowed to interact with each other at many places, where those already enjoying a better definition act as a sort of probe for the others, until the latter also acquire more precise definitions. The procedure is repeated throughout and gives rise to unavoidable repetitions, but I hope that the overall discourse will result in clarity. We begin our endeavor *in medias res*, i.e., discarding what would be generally held as the necessary premises, and jumping right into the middle of the questions. Such a procedure would be considered preposterous in any scientific paper, but, here the guiding logic must be different. We proceed following a partition of the matter, which offers a double advantage. First, it directly relates to the general framework that was the concern of the previous monograph, considered as the logical extension of the principles governing somatic and autonomic brain functions to the intellectual functions. Secondly, it approaches the problems by starting from a very general and more easily understandable overview of the eidetic function, the general problems of thought, and the dynamic aspects of eidetic processes. In the last two parts we confront conceptually more difficult problems, the very fundament of eidetic processes, i.e., the ideas, their formation, their nature, and the nature of memory and will. Finally, we turn to the fundamental origin of eidetics: language, conscience, and consciousness.

A final word to warn the reader about the way the matter is treated: the almost complete absence of experimental data of a neurophysiological nature forced us to use direct introspection as a substitute. Again, the purpose is not the systematic review of the philosophical tenets about mind, thought, etc., but rather to relate that which we partly know as a matter of fact with that which is partly accepted as assumptions or presumptions concerning the basic processes of central nervous system operations. The result contains many trivialities forming the background and much of the firm, supporting pillars for the thin spiderwork of theoretical constructions. The fact is particularly evident in Parts I and II, which deal with the very general problems of mind and thought, as general as can be afforded to draw the most general possible conclusions. However, in my opinion, one should not avoid apparent trivialities, because nothing is really so matter-of-fact that it can be blindly accepted: everything also needs to be logically cross-connected in a tight and comprehensive framework, the newest findings with the oldest and even trivial knowledge.

Part I

Problems of Mind

1 General Observations on Thought

The nervous system exhibits a progressive development from the simpler brains of amphibians, reptiles, and birds to the more perfected organ of man. The basic elements of the nervous structure —the neurons and their principal pathways of communication— do not seem to have changed substantially in the course of philogenesis, so that it would be difficult to argue that along with the progressive development of the structure there has also been a change in the more basic functional principles. However, it is clear that along with increased structural complexity new functional properties have emerged, such as language, which distinguish man's brain from those of other creatures. I should not like to make unwarranted assumptions about new principles of operation, when the anatomical structures of lower vertebrates and of higher mammals (and of man) exhibit the same general pattern of fundamental connections. However, properties such as the capacity for language represent something new and unique in the history of living beings. As far as the higher neural functions are concerned, we shall seek to provide them a general framework (such as is done in *PTN* regarding the somatic and autonomic functions) representing a comprehensive view of all nervous system activities within a unique, all-embracing framework of reference.

There is another reason for not assuming, at least from the start, different sets of principles for the different functions. In *PTN* we stress the fact that in every situation brain functions are based on the principle of a strictly unitary operation. The principle was maintained to hold for psychic functions as well, since there is no situation in life (thus, for our brain operations) in which the psyche is not involved, and in which the psyche in turn does not interfere with the somatic and autonomic operations. Moreover, one cannot deny that in driving man's life complex psychic processes overwhelm the simple paradigm of stimulus-response within the surroundings, as well as the demands of the autonomic system. However, somatic and autonomic functions in man are not performed in a substantially different way from that in other living beings; rather, they are subject to a higher control which allows modification of their course in accordance with broadly intellectual demands. As we have said, somatic and autonomic functions in man are performed in a more nearly perfect way than in other creatures, since they share something of the higher functions that man's brain has acquired.

Now, however, we should abandon the somewhat equivocal term "higher" in reference to functions not pertaining to the somatic or autonomic sphere. If we maintain that the cerebral function strictly preserves its unitary character in every situation, we can no longer set higher over against lower functions except when used in a merely colloquial sense (the term "higher" is not accepted by neuroscientists). Having abandoned the term higher, I should also like to use the term "function" more sparingly because it conveys the connotation of something "closed" or "finite," a definite set or goal-determind ensemble of neural operations, by no means the main characteristic of the operations that we are investigating. Let us use, instead, the more general term "processes," as those which in their ensemble endow our brain with the capacity of thinking; let us also use the terms thinking and reasoning as synonymous. We shall see that one cannot set sharp boundaries between these processes, including: a) the formation of ideas (sometimes with a very precise, other times with a hazier formulation), b) the recognition of the logicality of relationships among ideas or their constituting elements, c) the organization of thought, d) memory, e) analyzing and theorizing procedures, f) consciousness and awareness, and g) conscience and language. All of these processes could easily be divided into further subprocesses, but instead of multiplying the lines of analysis our task requires that we proceed in the opposite direction, unifying on the basis of common principles —by progressive abstraction of the essentials— the characteristic features of each of them.

Toward this purpose we first consider the internal language of our brain. The same term language may be used to signify three different phenomena: a) the brain "machine-language," represented by nerve impulses, coded for communication between nerve cells, b) the verbal language of our speech and thought, which we may call "external" and expressing c) an "internal" language —to be defined— which is the origin of the former one, but different from it. Language has been approached almost uniquely by psychologists and philosophers, not by neurophysiologists, since to approach language in terms of nerve impulses (the objects of neurophysiology) would require experimentation on human brain. This, however, is not our aim. For the most part in our endeavor we accept language as it is, in its whole and most general sense, because this property seems built largely upon the brain structure as the basis of all intellectual and psychic processes (see Part V). The components of language are like the bricks with which we give physical shape to the abstractions called "concepts," thus constructing the ideas and the sentences of thought. Only through language can we treat immateriality, which we cannot possibly refer to with a referential nod as we may do with material entities.

To understand the line of thought that we are following, I would like to focus attention on what appears to be the fundamental process at the level of brain circuits. In *PTN* we see that operations of the nervous system cannot rely on absolute measurements of any type —such as may be represented, for instance, by the amount of activity of individual neurons or centers— but rather that they rest on sets of "relative" measurements between large number of neurons or centers. This means that what matters is the pattern of distribution of the activity over many and different populations of neurons, and that every judgement, decision, or further processing of information is essentially based on sets of comparisons between individual measurements or whole patterns. Comparison is a broad concept, but fundamentally it requires measurements to be performed on the variables, as well as a measurement of the results of the comparisons. Operationally, the comparison itself may be performed with different methods, and with a somewhat different meaning for the different ways, but always within the same fundamental concept of "relating" different measurements to each other. The same term may indicate assessment of the simple difference between two activities (the latter in turn may be measured in one of the many possible ways; see *PTN*) or of their ratio. The fundamental utility of comparative measurements is that, when performed continuously, they provide a faithful description of the development of complex processes in which the neural activities are involved. However, when dealing with ideas and reasoning, simple measurements of cell activities do not appear suited for describing the complex significance of the eidetic procedures; therefore, better definitions are called for.

1.1 Thought

Even if thought is considered a unitary, indivisible process, it nonetheless remains extremely complex. To begin with, we should recall the distinction between two types: a) the logicomathematical thought, based either on actual mathematical elements or on elements reciprocally linked by similarly, easily definable relationships, for instance, general scientific thought; and b) thought which we may call "complex" because elements of a different type from the former are introduced in the reasoning, to a variable extent, and which belong to philosophical, ethical, religious, or social principles, those that we may call broadly philosophical. The relationships between the latter elements and the logico-mathematical ones seem sufficiently intuitive, but on closer examination they are not actually so apparent. The distinction is arbitrary and artificial and cannot be maintained for long, but its usefulness lies in having a gross semantic distinction which will help us through the early phases of the discussion. Nevertheless, we need to be more clear. This distinction suggests the existence of "rational" thought as somewhat in contradistinction to less rational, broadly philosophical thought. There is no sense in this, since it is obvious that broadly philosophical reasoning is as rational as is logicomathematical reasoning. This is not the real point. "Thinking rationally" is a redundant expression; thinking and rationalizing are synonymous, and both imply an ordered, logically connected sequence of ideas —in some cases we might even say necessary connections between the elements of thought. Although generally used in distinct senses, in their essence the two attributes logical and necessary, actually refer to more or less the same meaning, almost always in the case of logicomathematical reasoning and very often in the broadly philosophical.

If logic (and necessity) is the fundamental property of thought, the general construction of thought does not depend on its objects (for instance, logicomathematical or broadly philosophical), but only on the logicality of the relationships —the logical links— among the ideas constituting thought. Consequently, what really matters is how the logicality of the relationships is defined, i.e., how the congruence, or the coherence, between ideas can be determined. We have substituted the terms congruence and coherence for logicality and necessity, again treating these as synonymous. They are not synonymous, however, since coherence indicates a definite mathematical-statistical function, a measure of the degree of similarity between functions (for instance, power spectra) and therefore also logicality of sequences, such as in reasoning. On the other hand, congruence is more linked to the nature of two objects, such as the content of two ideas, in some way representing the degree to which they seem to "fit together," not as a true measure of similarity but rather as a judgement of logicality of construction, of kinship, of derivation, etc. I would like to stress again that both congruence and coherence are comparative measurements, still unknown to us regarding their material basis in terms of neural circuits, however, measurements that our brain can certainly perform easily even if not unequivocally, and based on still not universally accepted standards but bearing the heavy mark of subjectivity.

Although I have retained the term measurement in regard to congruence-coherence, what the brain seems actually to produce is no more than a "qualitative" judgement. In other words, from the comparative measurement stems a judgement of congruence-coherence to which we generally may attribute only a restricted range of degrees, such as good, poor, medium, and the like, although the measurement in principle span a very wide range of values. (This is not the only example in the nervous system —quite the contrary. Central neurons can produce activities at different levels of "intensity," but our judgements of intensity are rather poor for want of standards to which we may refer.) Thus, although congruence-coherence can certainly be measured within a sufficiently wide range of values, the judgement itself is necessarily poor, because the actual measurement of an absolute quantity has no reference for comparison within the brain (this problem is discussed in *PTN*). What we have thus far observed amounts to the view that thought is based upon sets of measurements performed on ideas with the purpose of establishing at every step the logical degree of their connections. However, I would not want to leave the impression that we are necessarily stuck with such poorly defined judgements of degrees of congruence-coherence. If we devoted ourselves to analyzing thought, we might be able to discover a better definition of the judgements for every thought and for whatever type of ideas. Moreover, we might be able to define a more or less arbitrary "internal standard" of reference, i.e., a unit of measure, so that the judgement might be given in terms of multiples of this basic, standard unit.

The problem of standard references is discussed in *PTN* in relation to variant and invariant functions. The widely differing conditions present in the brain when performing the measurements presupposes the existence of relativistic transformations in the state of the system. This is more than a mere academic speculation. According to this theory (certainly to be further substantiated by experimental evidence), none of the units for measuring the brain represents universally accepted standards; rather, they are only personal standards or, using physical terminology, units defined by "local metrics." The latter is subject to change, depending on the actual working conditions such as the presence of particular stimuli or of a constellation of stimuli, i.e., a "situation." If we accept this theory, there are measurements that maintain invariant after the transformation (having changed the reference unit accordingly) and other measurements that are state or situation dependent. The matter appears sufficiently simple, in theory, when the measurements are performed on elements of the physical world, justifying, on the one hand, the invariance of judgements based on comparisons between elements (of image, sound, touch, etc.) and on the other, the variability (state and situation dependence) of judgement about "absolute" quantities, i.e., those based on arbitrary, personal standards (see *PTN*).

But, in the case of our judgements on congruence-coherence the matter may not be so clear-cut and straightforward, and free from strong objections. If we accept our previous hypothesis that the measurement of congruence-coherence (a comparative measurement), defining the logicality of the links between concept-ideas, is ultimately subject to judgement (a measurement of absolute type), one might also wonder whether a logical thought could not appear illogical (or vice versa) due only to the local metrics —the personal standard of reference— i.e., in strict dependence upon the state of the judging system. Everyone can see that, once the factors which may influence the local metrics are defined (by hypothesis), we are led directly to question the existence of a "true" or unique rationality of thought processes in different situations. More important still, by admitting local metrics, we are necessarily led to question the rationality of judgement in different individuals —thereby raising the problem of the definition of their "normality." The point is that if it were possible to define the conditions which control local metrics, many ethical, religious, and juridical questions would have to be reconsidered.

1.2 Considerations on Thought Development

We concern ourselves now with the general mechanisms of thought development. Ideas are compared to one another, and differences or similarities defining congruence-coherence are measured. Considering the procedure as it is performed on sequences of ideas, one finds that it does not differ substantially from the principle of defining a correlation, i.e., measuring the extent to which knowledge or the definition of an object or a set accounts for the characteristics of another object or set. The different aspects of this comparison correspond to the different meanings attributed to our judgement on congruence-coherence from time to time, generally unconsciously, and determined solely by the object of thought. At every step, thought must thus recognize the result of the comparison just performed in order to select one of the (possible) alternatives for continuing thought. It cannot be maintained that reasoning always proceeds rigidly in this way. We shall see below that at least another, and quite different, procedure may be followed, one that depends on the purpose of reasoning itself.

What I would like to stress, is that there is a quite "natural" way (we shall return on the term "natural" below) imposed largely by the object of thought. This is because the elements subjected to judgement are so strongly, rationally, or necessarily connected to one another (for instance, the purely mathematical ones) that, in the extreme case, there may be no possibility of choice, i.e., no alternatives to select. In such an instance, thought follows preformed paths, and we tend to accept the existence of an "external" logic that mind merely recognizes and to which no alternatives are offered. The question is by no means trivial, and our answer rests on our personal standpoint toward science. Everyone would accept that —with the possible exception of a basic indeterminacy at the subatomic level— there is an order in nature, based upon relationships among the elements constituting it; this order follows a "natural" logic of things, represented by our physical laws. While no convincing arguments can be adduced against this, we must aknowledge that science is a product of man's brain, which has limitations in its structure and consequently in its possibilities of organization. Therefore, science, and the logic of the science as man discovers or recognizes it can only be that which man's brain is able to understand within the limits of its own powers, i.e., that which the brain's judgement accepts as true. This being the case, in the most general instance no class of elements of thought can appear to the human mind so perfectly and rigidly correlated that there can be no choice; therefore, the path that reasoning takes is not fully "predetermined." The choice is real; at every step it requires weighing alternative solutions and reveals at every moment its nature as a comparative selection (moreover, performed in a decided state of awareness).

We cannot expect the measurements of comparison always to be unquestionable. The intrinsically indeterminate character of the choice at every step of reasoning depends essentially on two factors. On the one hand, due to the individual's one preceding experience, he attributes a probability to each element of thought. This probability is not objectively or exactly determined but amounts to little more than a feeling or a sensation; every individual has his own personal measuring stick. On the other hand, each element of thought may give rise to a number of differentiated lines of reasoning that lead in different directions, with results that cannot always be foreseen. In this situation, the conscious or unconscious selection of the line to be followed is no longer made on the basis of a probability judgement, and one cannot predict which is the "more likely." Rather, there is a choice resting on some capacity of "seeing in advance" the course and the outcome of the line of thought, at least in a general outline. Knowledge seems to represent the deciding factor in this procedure. These two modalities, well differentiated in principle, cannot always be maintained as separate. Notwithstanding our ability in statistical prediction (in a broad sense) and our "foresight" capability, the actual outcome of the comparison between elements —representing the heuristic procedure of thought processes— cannot prevent any choice from being accompanied by the more or less conscious acceptance of a set of postulates and of consequential steps that have a bearing on thought which is not always evident beforehand. Thus, at some stage, the two procedures merge (though not on account of their intrinsic nature), so that the result of the reasoning is always affected by the predominance of probabilistic or determinative elements, even if not entirely and not precisely recognized.

1.3. Observations on Constancy and Development

The above outline of the mechanisms of thought views thought as beginning with logical connections among an ensemble of ideas (a logic measured in terms of personal metrics). A neurological operation (a comparative procedure) is seen as defining the congruence-coherence between the elements of thought, the ideas, and thus to represent the means for assessing the logicality of the whole line of reasoning. The same procedure can be assumed for the process of idea formation (this is considered more extensively below). Only some hints are presented here to show how the two processes, idea formation and reasoning —which we consider the kernel of the eidetic function— are founded on the same principles as the other neurological operations.

As in the case of other processes (and functions) considered in *PTN*, a major characteristic of ideas, as neural events, is their constancy, in this case of their "content," one of the factors ensuring constancy of behavior in widely varying conditions. Moreover, there is also an interindividual invariance of ideas content that ensures the maintenance of interindividual relations on a fairly constant basis. Constancy of the content —the essence, or meaning— of ideas is obviously the sine qua non for the processes of reasoning and represents the cornerstone of the thought structure. If it were otherwise, we would find ourselves in a real wonderland where the rules could change continuously (we shall see —next section— how the term "rule" or "law" used in connection with ideas is appropriate). As with other functions and processes of the nervous system, constancy of content of ideas seems to require a somewhat fixed structural and functional organization, even if the necessary circuits are probably considerably more complex than those of autonomic and somatic functions. On the other hand, as far as the ideas are concerned, a fixed structural and functional organization, while allowing a large degree of automatism of the basic eidetic process, at the same time ensures unconditional access to control by the will, and development in state of full awareness. This does not necessarily mean that we are fully conscious of every subtle step in the eidetic process —although based on comparative procedures (and on the related measurements) between elements— because we may sometimes take no heed of these procedures. However, we are always aware of the end result, which represents the basis of the final "judgement" on the process.

Constancy of content, or meaning, and fixed structural and functional organization (as the basis of constancy), make of each idea a kind of monad: a closed, irreducible, complex, self-sufficient system. If the elements in the eidetic system were always so rigidly fixed and each monadlike object strictly self-sufficient, there would also have to be an extended, suitable system of intercommunication between the structures; otherwise the measurement of congruence-coherence among the ideas as elements of thought might ultimately become altogether impossible. This may occur in a number of specific instances but not in the more general case. In the general case congruence-coherence measurement based on the comparation of the meaning of the contents allows the recognition of logical affinity or quasi-similarity of meanings when the ideas differ in only a few elements. This means that the monadlike idea is actually not a microcosm, and it implies that through addition or subtraction of elements the ideas may undergo some kind of development, either through increasing or decreasing complexity, such as a kind of dilatation or shrinking (for instance, from a more restricted or particular to a more general and comprehensive meaning, or vice versa, as is frequent in mathematical and physical reasoning). As we shall see, the eidetic mechanism of abstraction is based on special movement through the idea elements. In other cases, kinship or very close affinity may also be recognized between ideas not belonging to the same but to different classes or "regions" (see Sects. 3.4, 4.1).

The inherent possibility of some degree of automatism in eidetic processes, even if below the threshold of awareness, may allow development or transformation of ideas. In *PTN*, the development of brain states is also discussed from the point of view of "natural" motion throughout the phase space of the neural variables (although the meaning of the attribute "natural" as in statistical mechanics was not completely accepted). "Natural motion" is the physical concept by which a system develops in time (i.e., is subjected to transformation) without the intervention of external energy —in practice, the possible development of a closed system. The system's elements change their coordinate measurements in a purely statistical fashion; therefore, due to the mere statistical fluctuations the average energy content of the system is maintained as constant, and the system behaves as if moving on a surface of constant energy (on a hypersphere). The same, or analogous, principle (since some special form of "energy" may also here be in the foreground) could be assumed as being involved in the gradual development of thought. With this mechanism (if we may still use this term, since there would be no organization in this type of development) the direction of thought is not predictable in the general case, since the constituting elements are common to several lines of reasoning. A prediction on purely statistical bases could be achieved only through knowledge of the whole ensemble of elements in that set of ideas.

According to this view, the eidetic process in its whole would seem to represent some kind of "idle" motion without any precise purpose and therefore devoid of meaning, going on for the very reason that all the neurons are always active, even when not driven by other inner or outer events. All this would seem preposterous, but it is not. "Natural" motion throughout the ideas, although not guided and essentially purposeless, does not necessarily lead to altogether chaotic or irrational thought. Simply, natural motion would randomly "present" ideas to the structures effecting the comparison, and if the latter are operating consciously, a high degree of rationality would still be ensured by the fact that the judgement on the results of the comparisons is a rational process and cannot be left to chance. Nevertheless, the whole procedure might become unbearably lengthy. (Surely, however, we cannot attribute to this "idle" motion —the mind's wandering about without a previously defined goal— the same meaning colloquially given to idle, or leisurely thinking about some defined object, as thought is addressed in a definite direction.) However, even if at first sight natural motion seems a poorly effective mechanism for thinking, we shall see below that under certain circumstances it can reveal itself to be as efficient as other types. However, before we may judge on this matter, we need more insight into the mechanisms of idea formation (how the elements of the ideas are assembled) and the mechanisms of idea development (or transformation), i.e., on idea motion.

1.4 Ideas as Laws

We used the terms rule and law in connection with the constancy of concept-ideas. The relationship here is much more than an analogy. It is my view that an idea does not differ from the expression of a rule or law, and that it can be represented and treated as such. In the most general case, the elements composing an idea are assembled following well-defined procedures, for instance: given A and B, then C. Thought also results from assembling its elements —the ideas— in a very similar way: given the ideas A and B, then, C. This becomes clearer if we expand and rephrase the sentence: since the elements A and B can be put into some specific relationship to each other, then, C represents the law, or rule, relating a and b (or some other logically connected sentence). This is nothing else than the way in which mathematical and physical laws are formulated, using knowledge of the elements involved. This way of looking at ideas, moreover, entails that when the relationships among the elements are changed (for instance, their relative measurements), the idea too is changed. This means not only that the comparative procedure (which measures the relative values of the elements) is at the basis of the formation of the idea, but also that the same set of elements, provided their relationships are changed, may give rise to ideas that are sometimes altogether different. It may seem awkward to accept beforehand that "every" idea can be formulated following this (or a similar) procedure, i.e., that it may always be represented as a law (this question, far from being settled, is considered below).

However, one aspect of the problem must be mentioned here (and treated further, below). In the formulation of a mathematical or physical law the elements are all laid down simultaneously, so that we can understand its meaning at a single glance. This raises a far reaching question concerning the eidetic process. The matter is not at all trivial, if we only think that the eidetic process takes place in a material substrate constituting the neuronal network, where only time sequences of events can be subjected to comparison or can be cause-effect correlated. Physically speaking, it would be dubious to assume the simultaneity of nerve events, due to the different, finite times required by the different events to reach the processing structure. In *PTN* much attention is devoted to this fact due to the close analogy with the situations considered by general relativity, which claims the impossibility of demonstrating simultaneity of events in the physical world, where only finite, though very large, velocities are involved. In *PTN* we advance the reasonably well-founded hypothesis that brain function may be viewed better, and explained more easily, within a relativistic reference framework. The view stemmed from the observation that the characteristic unit of time measurement within the nervous system, to which every nervous event (motion) should be referred, provides no fixed reference for the latter but represents a state-dependent variable, a "personal" time (related to the system's *Eigenzeit*). Thus, the local metrics, which, as discussed above, provides us with personal "yardsticks" for measuring internal and external variables, is involved in the brain's time as well: a typical, clear-cut general relativity situation.

Thus, the problem represented by the necessarily simultaneous presence of all the elements of an idea —so that they can be recognized as elements contributing to the formulation of a law, and therefore of an idea— must be solved in terms of brain structure and organization. Before we can do this, however, we must consider whether the idea does not result, rather, from a slow process of construction, by successively recruiting the elements, adding one to another only after their congruence has been assessed, and it thus takes shape stepwise. This instance seems in contrast with the more frequent flashlike appearance of an idea, being "lit" by the simultaneous presence of all its elements, perhaps catalized by one of them. This does not refute the view that the idea was there and only flashed when all the elements had been assembled with the right coherence and in the right order. It is clear that the procedure through the structural organization in the two instances must be different, but it is also clear that they must converge in the final, characteristic stage of the idea coming to life.

1.5 Natural Eidetic Motion

We can certainly conceive different ways of building an idea, and not require the simultaneity of all its elements from the very beginning of the process. Looking into ourselves and trying to analyze the mental procedures leading to new ideas, we may find occurrences of both the opposed ways of idea formation: sometimes one is struck by a sudden idea, and sometimes one arrives at an idea in a piecemeal and painstaking way. Leaving aside for the moment such flashes, the neurological process underlying idea formation can be represented by a kind of sweeping across the ensemble of available elements —a current elements' "phase space"— recruiting only those that exhibit a congruence-coherence in terms of their intrinsic meaning (as measured by a comparative procedure). Here, again, the type of nervous motion that we are confronted with could have a "natural" character in the sense used in statistical mechanics . But the latter motion cannot be accepted in the simple form of an exploratory mechanism of the phase space in a closed system (left to itself, or isolated). As maintained in *PTN*, the process does not seem to be left to chance, as the statistical mechanics hypothesis requires. Rather, the new ideas built by the exploring mind seem always to result from a preceding reasoning, charging the rational process with the task of finding the right elements in the phase space. Regardless of our view on this still debatable point, certainly we must concede that the procedure of adding the elements to one another would be more easily conceived, and appear more likely, if the elements of phase space —in practice phase space itself— constituted some kind of memory, or *the* memory, from which one can select and retrieve the components at will. Thus, the problems related to eidetic processes are intertwined with one of the more obscure and often-discussed functional properties of mind, memory.

Questions related to memory we shall not discuss presently. As noted above, if we accept that building an idea requires recruiting the elements in a given order according to the criterion of their reciprocal congruence-coherence, it is clear that we are envisioning a process identical to, or at least not substantially different from, the process of reasoning; the basic eidetic procedure would be the same throughout, from ideas to thought. The continuous activity of brain neurons would appear to support the hypothesis of a continuous "natural" eidetic motion, in the sense outlined above. Again, however, we reject this as an orderless motion, one which in the general case is poorly effectual, and we suggest instead a "perpetual" eidetic motion supported, indeed, by continuous activity, but of highly organized, rationality-based subsystems. This provides for: a) grouping the elements on the basis of their congruence-coherence, thus making it easier to build ideas; b) performing the transformation of ideas by adding or substituting elements or by altering their reciprocal relationships; and c) ordering the ideas and building sets out of them, thus promoting thought, a process developing in time. Although it is reasonable to accept that the steadily ongoing activity of the neurons represents the neurological substrate of all the continuous processes, we have no actual evidence that ongoing activity alone represents a sufficient support for the mental functions as well. As already indicated, eidetic function and memory are closely related processes, perhaps so tightly linked as to be two aspects of the same function. Thus, the eidetic process would not require continuous motion if it could count upon constantly present (for example, memory-stored) elements and ideas (which seems intuitively evident, even if not clearly demonstrated). With these considerations we seem to have reached the core of the problem, but this is not the proper time to tackle these complex questions.

1.6 *Geschlossenheit* of the Eidetic Constructions

We have maintained that the fundamental and normal procedure in mental processes is comparison, as a measurement of congruence-coherence, according to one of the possible ways of operation of the neurons. This has led us to the view that the two aspects of the eidetic function, i.e., the formation of ideas and the organization of thought, are quasi-identical and have the same basic significance in recognizing an inherent *Geschlossenheit* of the elements constituting it. Recognition of the greatest degree of congruence-coherence among the elements of an idea or among the ideas as elements of a line of reasoning is seeing them as constituting a unitary, closed system of their own (i.e., *Geschlossenheit*). Each idea or thought enjoys a certain independence from the thinking mind itself, since the *Geschlossenheit* is determined by the essence of their content, i.e., by their nature, and represents as such the basis of interindividual invariance (being shared possibly by many). The fact that the same idea or thought in different individuals might exhibit some difference is immaterial, provided the fundamental significance of that mental construction remains essentially unaltered. As we shall see, the differences may reside in some of the elements, as variables characteristic of each individual.

The quasi-identity of two ideas having fundamentally the same content hints at the possibility of progressive mutation of the ideas. Therefore, to the concept of fundamental invariance of the substantial core of an idea or thought we shall have to add the concept of "contacts" between ideas (notwithstanding their *Geschlossenheit*) and that of graded transitions between them that may easily make them a self-sufficient "class." The relative independence of the mind's constructions from the thinking mind itself, as a result of interindividual invariants (see below) is certainly a fascinating metaphysical concept, but is beyond the scope of neurophysiological investigation of mental processes. Our purpose is to discover (if possible) how a metaphysical object may arise from a neurological operation such as the exchange of impulses between neurons and the organized interactions of nerve circuits. However, what I have described as the kernel of our problem actually is still hidden beneath many concentric layers, as onion skins that need to be patiently peeled off one another.

1.7 Physical Implementation of Ideas-Laws

A characteristic feature of analogue computers is that they solve mathematical equations by "embodying" the formulation (equivalent to a system of relationships or a law) in an electronic circuit. Today, the use of analogue computation is restricted to a number of special purposes only, since they require a different philosophy of approach, and are generally more expensive than the more widely used digital computation. In analogue computation, particularly in real-time simulation of physical situations, the problems can be more easily, and almost visually, solved by the method of "implicit" implementation —particularly in the case of higher order differentials and trigonometric equations— through rather simple circuits. Conceptually the latter is based on a number of feedback loops. There is a close, though purely formal, similarity between physical analogue circuits and eidetic circuits. The former implicitly embodies the formula, in which the ensemble of relationships among the elements entering the formulation (the law) may be said to be "realized" at the moment the feedback loops are closed. It is in the latter where the idea appears, representing the law —the relationships among its constituting elements.

As we have observed, the idea may appear in the eidetic circuit at the moment the *Geschlossenheit* of the elements (defined by the reciprocal relationships making them into a unity) is recognized, i.e., when their reciprocal congruence-coherence shows how necessary their linkage is. In *PTN*, the significance of the feedback organization of the brain structure is emphasized in the fact that the feedback represents the substrate of a functional unity. In the maze of the intertwined feedback circuits, different elements, each with its own individual meaning, interact to produce new subunities with an intrinsic *Geschlossenheit* of the elements comprising it. Thus, an idea appears as a unity in the same way in which an intrinsic unity appears in an analogue computer feedback circuit in which a law is implicitly implemented, when the circuit's elements are allowed to interact with one another.

We must stress here, however, that the conceptual similarity —with the electronic analogous implicit embodiment of physical laws and mathematical formulas— by no means implies the actual material similarity between brain and computer circuits. The only point of similarity between the two embodiments (and I believe in this they are identical) lies in the fact that the formation of the idea may also be founded on feedback processes —thus, on a "principle" of construction as well— in a closed structure representing, in its completeness and unity, the whole set of relationships among the constituting elements: a physical law, on the one hand, and a mental law or idea, on the other.

The preceding considerations, besides suggesting the existence of material structures of the eidetic function, perhaps "fundamental unities" through which our mental processes are built, might also open the door to an experimental approach in a neurophysiological sense to the study of eidetic activity. This raises a series of questions concerning: a) the possibility of genetic inheritance of abstract (mental) structures embodied in neural circuits; b) the individual ability to construct ever new mental structures; c) the consequent training in mental processes; and d) the stability of the neural circuits where ideas are embodied. In addition, not the least of these are the fundamental questions posed at the beginning, i.e., whether the possibility of embodiment in neural matter may be in some way limited to only a few types of ideas and whether our reasoning may itself be constrained by certain limits or predetermined along certain paths.

1.8 Eidetic Operator

In this section we examine the hypothesis that the possible transformation of ideas and the development of thought may be the outcome of the interaction of an "eidetic operator" with the elements of the idea or with the ideas of a chain of reasoning. The concept of operator as a determinant factor in transformation does not seem as straightforward in the case of eidetic function as it may in the case of other nervous functions. However, this is only because the eidetic function has seldom been looked at from the neurophysiological viewpoint and never with the eyes of the physicist. As we shall discuss below, the level of abstraction with the eidetic function is surely much different from the rest of the nervous functions, and therefore the concepts of physics that we want to apply to it must also be duly abstracted to the same level. In doing this, we follow the principle that the laws of physics rule every activity of the brain, at all levels, from autonomic and somatic to that subserving the intellectual (and spiritual) functions.

The concept of operator is closely linked to that of "state" and is justified in the context of the eidetic function only if the latter can be considered in terms of a continuous transformation in the state of the eidetic subsystem (if not of the brain system as a whole). Regarding the idea, the transformation may consist in the development of its content, in it progressively taking on newer, somewhat different contents. Regarding reasoning, the transformation occurs at every step where a choice is made between alternatives and represents the progression of thought towards its goal. We have repeatedly denied that the "motion" of mental processes can be represented by so-called "natural" motion (in the sense used in statistical mechanics) although no proof has been offered for rejecting an hypothesis that could certainly simplify some of our discussion. Besides the arguments advanced in the preceding sections, one reason for the rejection is that the nervous system is not closed in a thermodynamic sense (see *PTN*), which barres this interpretation. However, as far as brain is concerned, we have also seen that the constant energy requirement of a closed physical system may be circumvented if the energy involved is not the metabolic or biochemical energy substrate of cell life. A "nervous energy" inherent in motion parameters could be maintained as stable, provided sufficient, albeit widely varying metabolic energy is continuously supplied. However, we have also questioned the very existence of an energy constancy, because brain activity consists of an endless sequence of "states," each continuously transforming into another —a continuous motion, or development, not determined by a chaotic process but resulting from operator-state interactions aiming at a specified goal. (We shall see that some form of "eidetic energy" may be envisaged, accounting for exchanges between eidetic spaces and for "attraction" between eidetic objects.)

If we accept the arguments advanced in *PTN*, we may start from the assumption that every step of thought is a process somehow guided by an operator, the "eidetic operator." However, we should remember that according to our theory the operators in the nervous system are not represented merely by a physical parameter (such as a dynamic variable in quantum mechanics) but by the activity of nervous structures not easily reducible to one single and simple variable. The crucial point is that thought development does not appear to be left to chance, i.e., within it there is a rationale of some sort, and is a typically voluntary and thus purposeful act. That thought has within itself a rationale belongs to the very essence of thought; however, that will and purposefulness are also part of its essence may be a matter of definition, although this is common belief. Will and purposefulness imply full awareness throughout the process, and we may say that awareness befits the nature of thought, there being no evidence that the continuous choice operated at each step of thought development may occur outside awareness.

That some mental processes may continue also in subconscious conditions, ought to be accepted on account of the continuous activity of all nervous structures, thus presumably also of those involved in eidetic processes, but there is no demonstration that a really organized, purposeful thought ever resulted from the activity in non-conscious states. This still leaves open the question as to the physiological significance (if any) of the unconscious activity of brain, (some hypotheses will be offered in later sections). Assuming that at least one operator (but probably more) is guiding the reasoning process, we should still maintain a distinction between the different steps of thought. There is first the process of comparison for determining the degree of congruence-coherence among the elements constituting the ideas. Then comes the judgement of congruence-coherence among ideas; this is subjective (and possibly state dependent) although based on the appraisal of true comparative measurements. The final step is the search (an exploration, or sweeping) across phase space for new ideas to submit to comparison. It is clear that if memory were involved —representing the actual phase space of elements and ideas— an eidetic operator would have many characteristics in common with a memory-reading operator.

The concept of operator is borrowed from quantum physics, where the operators are dynamic variables. The operator is applied to a state vector (a "ket" vector) in either of two situations: a fixed (time-invariant) operator and a moving (time-dependent) state ket vector (the Schrödinger situation) or a time-varying operator and a fixed state ket vector (the Heisenberg situation). The general formula —operator times state ket— can be assumed for the nervous system to account for the development of the nervous system state (in *PTN* considered as true motion, although not in physical space), in terms of series of state transformations as in quantum mechanics. Further extending these concepts to eidetic processes may help us to account not merely for the large "jumps" in the content of ideas (here not a straightforward concept) but principally for the minute transformations, such as the gradual modification of significance and essence, thus a gradual transition from one to another idea (and for the development of thought). The operator effecting these minute-step transitions would closely resemble a "unit" operator (as in quantum mechanics), or a "contact" operator (as in classical mechanics). Accepting the minute-step procedure is probably the only way to accept the concept of idea development, as the one ensuring at every step the best congruence-coherence between the preceding and the following idea, since it would "work" on one element (or a few) at a time. Only in this specific sense could we call it the "natural" transformation, and the corresponding operator the "natural" eidetic operator.

We have not yet raised the problem of whether the eidetic operator is fixed or time dependent, with the corresponding alternative of time-dependent or fixed state ket. We have only hinted up to now that in its *Geschlossenheit* the idea is like a closed subsystem, and by that —even if the analogy with thermodynamic closed systems is not fully applicable— in some way time independent too. Accepting this argument, the operator could only be time dependent, as is assumed to be more probable in the theory of *PTN* on account of the general situation of the brain functions. Of course, at this stage we dare not to assert that the time-dependent (or "moving") eidetic operator is the same one which performs unit state transformations in somatic and autonomic subsystems, therefore, the question must be left open. Progressing through infinitesimal steps renders more difficult an overall sequence of chaotic transformations but does not entail by itself that the direction is pre-determined.

The idea is a unity, a closed system only by virtue of the *Geschlossenheit* of its many elements, and the fixed (time-independent) state of the idea is defined by the invariance of the elements. The natural eidetic operator, according to our definition, does not involve simultaneously the whole set of elements of an idea: the microstep transformation of the idea is best accounted for by a "selective" operation on one or a few elements. It becomes almost obvious to accept that the operator itself may also select the direction for the best congruence-coherence, or affinity, between the result of transformation and the original idea, by selecting the entering elements. If, at first sight, the concept of fixed state and moving operator seems to have been unduly stretched to account also for the transformations of an eidetic subsystem obtained by operating only on a few elements of it, at a second thought the procedure seems justified. In quantum physics each state vector (or ket) can be considered as resulting from the linear superposition of an undefined number of other vectors, each one contributing with its own weight —represented by a coefficient— to determine the resultant ket. We can transpose the same concept to an "idea state ket" as defined by the weight, or coefficient, of its elements. If one or more elements are changed, the weights (thus the coefficients) are changed accordingly, and the resultant state ket is transformed, in quantum physics as well as in the eidetic system.

The outcome of the operation evolving through micro-step transformations cannot be an indefinite concept or an illogical reasoning. Even if the procedure is represented by chaotic trial-and-error microstep sequences, the resultant micro-step motion of the ideas, being sanctioned by a judgement on congruence-coherence, could never be confined on a surface of constant energy (as a purely chaotic sequence of events necessarily would). Here, the hypothesis is that the development of ideas and the progression of thought follow well-defined thermodynamic gradients of eidetic energy, an assumption inconsistent with constant energy surfaces since the motion steps would exhibit trends and become time dependent, thereby crossing different energy surfaces. This amounts to assuming that the operator itself, by the very fact of being an eidetic operator, "knows" the direction in which to search for the more congruent element ideas,and may find them quasiautomatically, or "naturally."

We must recognize that there may be different possible ways of idea and thought development. If we accept the principle borrowed from physics that ideas and thought are determined by the weights (or coefficients) of their elements, it can easily be conceived that some of the elements, because of their connected weights, may in some way "hook" the operator onto themselves, thus becoming the direction determinants of the transformation. Moreover, the transformation operator can be will guided, which seems obvious in eidetic matters (see Sect. 5.4). Finally, one ought not to assume that the different possible ways of idea and thought development are mutually exclusive; rather, we are inclined to think that they are all posssible, only representing the different procedures used from time to time, according to the needs of the development.

1.8.1 Properties and Structure of the Eidetic Operator

Several more aspects need yet to be examined. There are a number of requirements for the neurological structure of an eidetic operator. Considering only the neurologically acceptable hypotheses, in the eidetic processes there are aspects that seem at first sight to be of secondary relevance, but that offer a key for the unitary explanation of mental events. Two situations seem particularly important due to their bearing on our general theory of nervous functions. a) When developing a line of reasoning, the line can be interrupted at any moment and later resumed, even after a long time period, at the point at which it had been abandoned. Thus, there is no need to retrace the whole path which led to that point. b) A closely related but different situation, occurs when the conclusion of a given line of thought can be reached simply by "jumping" to it, without going through the whole sequence of logical steps, that is, short-circuiting a sequence of step transformations. While in the first instance the reasoning is followed up to that point through logically linked steps (another explanation is easily conceived, such as the existence of memory traces), in the second instance the intermediate steps are neglected, and the conclusion of the reasoning is "seen in advance" (a given extent of capability for foreseeing was suggested above as one of the ways in which we conduct reasoning).

It is all too easy to assume that resuming a line of reasoning after it had been abandoned for some time is merely a question of memory. The stepwise following of a thought, when at each step a judgement on congruence-coherence is given, might also be seen as having the side effect of fixing in some kind of memory the whole procedure, so that its resumption would only be a matter of retrieval. The capability for seeing in advance, on the other hand, might appear as something of a supernatural property, not very different from the assumption of a "prescience" —an absurdity, unless one accepts that at least some of mind's content can be inherited, or imprinted very early, including matters in the whole range of scientific and of broadly philosophical knowledge. Since one cannot accept this the underlying mechanisms must be different. Nonetheless, in some way memory may be involved in both the phenomena that we have been considering. Certainly, one cannot dismiss the feeling that one makes recourse to memory too easily, albeit quite reasonably, for explaining mental and nonmental functions as well. One cannot deny the existence of some kind of memory, because there are phenomena of apparent recall. However, that the memory of ideas and thoughts, of events, of images and other sensory experiences might be supported by some type of structural or chemical stable coding, although still a common belief, is also not easy to accept due to the lack of clear-cut, direct, and convincing evidence. The natural desire for symmetry, in this case with genetic codes, facilitates uncritical acceptance of unjustified concepts, also frequently originating in the careless use of a single term for different phenomena, as perhaps in the case of memory. On the other hand, the perceived difficulties inherent in the structural and chemical hypothesis have already fostered hypotheses of a different type, essentially based upon recurrent (read: continuously recirculating) activity in feedback loops (as is now generally accepted, accounting at most for very short-term memory). Again, this results from the uncritical presumption that memory can exist only in the form of something "stable." (We discuss below, Sect. 7.1, the significance of stability in relation to memory.)

Let us return to our task of finding those principles of the structural organization that may account for the particular properties of the eidetic operator. The basic arrangement of neurons in information-recirculating circuits has certainly been overestimated regarding their possibility for memorizing neural events, even for the short periods of a few seconds. On the one hand, one has overlooked the fundamental property of feedback, that is, the continuous interplay between every signal introduced into the circuit and the activity of the elements within it. On the other hand, the extremely wide system of cross-connections among all nervous structures has been underestimated. One cannot possibly conceive that in any neural structure an event may be maintained unaltered, on account of the fact that all the neurons are continuously active, and consequently the inputs to these circuits are also always active —although not necessarily with the same repeated patterned sequence— and therefore their effects interfere continually with the newly incoming inputs. (We shall see that every reasonable theory about neural memory must take heed of these facts.)

When these considerations are applied to eidetic circuits and the general characteristics of operator-state interactions, one is forced to recognize that the result of the operation may also modify the way in which the operator operates, which means that the operator itself, in turn, is modified. The hypothesis appears more than likely if one only considers that the number of inputs to every neuron may be very large and from very different origins, generally not all simultaneously active, so that every particular pattern of neuron output depends on the momentary "constellation" of the inputs and its immediately preceding history; no neuron (and no circuit), can therefore continue to do exactly the same thing throughout. On these grounds, there is no reasons for expecting invariance of the operators, —except in some fundamental properties— since these are being continually updated by the effects of their operation, by other supervening inputs, as well as by the remaining results of their preceding activity (i.e., their history).

The general hypothesis here is that a suitable stimulus sets the operator to work on elements, and thus ideas are developed and thought is in progress. In turn, the eidetic abstract structures —ideas and thought— modify the operator, and the latter again develops more ideas, but at this time, following somewhat different pathway and procedure, both because of the different working conditions and the intervening change in itself (figuratively speaking, one might say that the operator has become "wiser"). As noted above, this sequence can be cut short voluntarily and later resumed at the point of interruption. Upon resuming the same line of thought with the stimulus which formerly set it running, the whole sequence of interactions between operator and ideas would no longer be required because the operator now functioning has the form determined by the previous operation. This explanation, however, does not seem to avoid the necessity of some kind of memory process, one that would involve not the sequence of events but the structure of the operator, less probably the conditions of the last operation.

Nevertheless, we maintain that still there is no need to assume storage in some memory of the operator's structure or of the exact conditions of previous operations. We propose, simply, that what is involved is a changed dynamics for evoking or constructing the operator, based upon an improved or different "nonspecific" performance capacity of the neurons and therefore of the eidetic circuits. Once more, the ground for the hypothesis is represented by the general properties of brain structure organization, and precisely those involved in a phenomenon akin to learning, with still poorly understood bases. As is generally conceived by modern theories on learning, each specific action —a stimulus or a set of stimuli— leaves behind traces, i.e., a "memory" (reinforced if repetition takes place until a stable representation is obtained). However, this general memory-related paradigm does not necessarily cover all learning processes. For instance, it is well known to experimental psychologists that test subjects (and animals) may learn "experimental sets," i.e., they learn not specific items or exactly defined situations but simply how to behave in situations only fairly similar as to general conditions, such as the necessity of finding correct responses when the stimuli and the situations are changed. This learning is nonspecific and is represented in an improvement (a facilitation?) of performance not based on strict identity of conditions but only on generic analogies between situations. Since this does not conform to the classical learning paradigm, it may not even be considered to represent learning at all, although surely names may be found for it, for instance, "generalization."

The basis of the improved performance may rest, more simply, on threshold changes, on increased velocities of processing, perhaps on simultaneity of processing through parallel channels, or on possible relativistic effects involving time and space parameters of the neural activity. With this, we by no means need to postulate structural changes in synapses and circuits; certainly such changes could also account for the opening of new or unused synapses not necessarily as specific memory substrate —rather, as structural support of the improved capability for idea formation and eidetic processes in general. It seems much more reasonable to interpret the experimental evidence of an increased number of synaptic contacts (though still not clearly quantified and unequivocally correlated) in the sense of nonspecific improvement in neurons performance, and thus also of mental processes, than in the sense of specific traces for every item to be memorized. If one accepts this increased capability for developing eidetic processes, the operator's capability for "jumping to conclusions" also becomes more easily conceivable since it would entail avoiding part of the —sometimes lengthy— stepwise development of reasoning. No one can deny that there is a training in mental activities, a progressively increasing performance in different mental tasks induced by the continuous use of the eidetic structures (as if continuously submitted to similar experimental sets), not unlike training in those motor exercises which require improvement of neuromotor coordination and organization more than increased strength due to growth of muscular masses.

One of the reasons why the brain of the human infant seems only slowly to reach maturation (not simply of histological and chemical structures but especially of their organization) is the necessity of becoming master of the fundamental eidetic processes peculiar to humans, with which it may start its mental training. We still do not know, nor can we foresee, to what degree of perfection the eidetic training may eventually lead the human mind. However, we know how the process begins, i.e., with the training that leads to learning experimental sets, which in lower species represents the highest level of performance reached, whereas in man this is merely the beginning of a continuous development of the intrinsic, peculiar eidetic potentials that only aging, disease, or death can truncate (see also the "Appendix" of *PTN*).

There is at least one other possible explanation, not an alternative to that just discussed, but complementary to it. We have suggested that the operator itself in some way must know the direction in which to seek the more congruent ideas, finding them quasiautomatically. This amounts to the assumption that the operator's function in selecting the ideas as elements for constructing thought is backed up by some preknowledge of the scheme of reasoning (which would be not at all unreasonable). This point will be taken up again below, Sect. 2.1.1, but it is obvious that an aim *is* intrinsic in every reasoning procedure, and that each step of the latter must be subjected to comparison with the aim outlined earlier toward ascertaining correspondence with it (i.e., that the reasoning is not going astray). The general character of this situation is one of the most common in the central nervous system: the whole control of motion, all the automatisms in the somatic as well as in the autonomic systems, and a good many behavioral reactions are controlled according to this principle, organized in matter as a continual feedback exchange of information between the structures from which the procedure originated and the regulating ones.

It seems extremely simple to envisage an eidetic operator as directly represented by the last step reached, or by a function of it, which would itself become the agent for the next step, under the continuous control of the originally outlined aim. What this hypothesis suggests, is that reasoning proceeds along "consequentially chained" steps, according to a generically intended causality mechanism. In this way, the judgements on congruence-coherence with the original aim at every step are still not indispensable because no step can be determined uniquely by an antecedent (if not exceptionally, and only in a mathematical-like context; we deal below with the consequential, cause-effect relationships, Sects. 3.3, 5.1). As to the question of remembering the last step previously reached upon resuming a reasoning, the arguments concerning the nonspecific mental improvement may become more precise. We need only to consider that the link between the stimulus that originally started the reasoning (obviously strictly connected, or perhaps directly identifiable with the aim) and the last step reached might consist simply of a facilitation of their association, reinforced by the fact that a judgement on congruence-coherence has already sanctioned their strict relationship, and no further step has intervened. However, there may be more. As we shall see in Sect. 2.1.1, the original aim and the last reasoning step, after their coherence has been determined, may become part of a knowledge which, albeit incompletely developed, remains until that last-reached step the only knowledge available about a particular item and therefore an abstract object retained by mind.

By talking of operator, I do not wish to convey the impression that we are dealing necessarily with a separate entity operating "within" the eidetic structure. In quantum physics, from which the analogy is derived, the operator is a dynamic variable, i.e., part and parcel of the same system. Likewise, in the eidetic context the operator may be any of the parameters of the eidetic motion. Therefore, proceeding further along the same line of hypotheses, we can identify the operator with the scheme from which a particular reasoning originated. This assumption is not substantially different from the preceding one, for the operator acts as a logical link in the thought process, and it does not matter whether it intervenes as the act of connecting the original stimulus to the ideas, or the ideas to one another in the consequential chain. In both instances, before operating on one idea, the operator has already interacted with the original scheme of the thought (since the operator must know it) and with a larger complex made up of the original scheme and of the chained ideas resulting from the preceding operator activity. (We shall see that this picture becomes more easily understandable when we identify this original scheme with the "concept.")

The point that I want to make now, by viewing the dynamics of the operator as the interacting link between original scheme and ideas, is that a judgement on congruence-coherence is no longer required, since at every step the presence of the original scheme (let us call it the concept), having acquired progressively better definition in the run, would prevent the reasoning from actually going astray. From another point of view, the eidetic operator appears as a probe through which the ideas acquire their specific sense within the context of a specific reasoning. We have just claimed that the improvement of mental capacity basically is a nonspecific phenomenon, not necessarily founded on the specific content of the mental process. However, this is only a minor aspect of the improvement that can be reached by continuously operating the eidetic structures; it represents only the beginning of a process leading to the development of the "power of abstraction." The basis of the whole eidetic function, in the very general sense, is the capability for abstracting the invariant out of the contingent, the essentials out of apparent properties; it is the power which specifically renders reasoning possible in purely abstract, not matter-bound terms. Once again it should be stressed that the procedure is based on comparison, but when the abstracting power is involved, the same procedure becomes more complex because the comparative terms for each particular idea must be found in a specific region of knowledge, requiring a kind of "sublimation" of the simple abstraction involved in learning an experimental set. (We shall develop this argument below.)

1.8.2 Interactions Between Operator and Eidetic State

As suggested above, accepting the conceptual scheme of a development of the eidetic processes through the interaction between ideas and operators also entails acceptance of the hypothesis that the eidetic event (in the extreme, also a single idea) can be considered as a "state" of the concerned nervous structures; thus, reasoning would consist of a sequence of states of the eidetic subsystems. In *PTN*, the theory of nervous states is discussed in detail, the term state indicating a state of motion (in philosophical and in classical dynamics senses) of the nervous system as in any other physical system. A state comparable to one of accelerated motion was assumed whenever the brain is engrossed in activities requiring attention to "interesting" problems, and for the latter a general relativity framework was also proposed. This relativistic situation seems to characterize eidetic processes, and the latter, even if not concerning in turn other brain functions (a very unlikely occurrence), ought to be considered as representing a state of accelerated motion of the supporting structures, affecting also the Minkowskian time coordinate together with the Lagrangian position coordinates. The accelerated motion of the eidetic subsystems creates a relativistic situation when other brain subsystems are in a different state (for instance, a time-independent, constant-velocity, inertialike motion).

As is seen in *PTN*, the existence of relativistic effects in the brain, even if not yet supported by sufficiently extensive experimental evidence, is a well-founded hypothesis. Up to now, brain relativistic effects have been detected through judgements on time intervals only, for want of suitable tests for relativistic changes in spatial parameters. Since a state represents a reference framework, as a system of coordinates, one must postulate that the transformations of state of the eidetic system also represent a succession of reference coordinate changes. Furthermore, one must take heed of the fact that since the object of the eidetic processes may offer different degrees of interest and of intellectual engagement from time to time, the motion of the supporting structure may accordingly be more or less accelerated and even become "inertial" in the extreme cases. The brain is continuously engaged in performing functions, both somatic and autonomic. Correspondingly, its general state is changed by operators belonging to the so-called diffuse projecting systems, not necessarily related to eidetic operators. However, a general physiological state change, such as the transition from arousal to sleep or vice versa, would strongly influence the eidetic processes as well, and therefore the eidetic operators. Actually, in order to influence or even impair the eidetic functions, drastic changes in general state are not necessary. For instance, pain and other emotional factors regularly interfere with the course of every type of reasoning. Although in these interferences there may also be a strong component related to attention (distraction or intensification), the general effect on thought development certainly cannot be attributed exclusively to attention changes.

The reverse also is true. We have seen in *PTN* that the somatic and autonomic functions (moving, sensing, feeding, mating, etc.), although presenting the same substantial characters in man and animals, are perfomed by man in a different way due to the fact that are executed by nervous structures strongly influenced also by the eidetic functions. As a matter of fact, the corresponding anatomical substrates are organized following the general principle of brain growth in philogenetical evolution, i.e., the growing together of philogenetically older and newer structures. This principle endows the older structures with some of the properties of the newer ones, and the latter with properties of the former, so that all the brain functions of the somatic and autonomic spheres are performed by man in a different, and in some sense more "perfected" way than by any lower species, because partaking at the same time of eidetic functions. Since the eidetic process is likely to influence the general state of the brain through enhancement or depression and that of attention (due to the interest aroused), and the latter in turn interferes with the eidetic development, we should not wonder that the operational procedures of the operators may be changed as well. According to our hypothesis, at different times the same stimulus would not succeed setting in the same operator to work, but rather a different operator because the system would no longer be able to produce the previous one due to the changed nonspecific performance capacities resulting from the changed general state affecting the eidetic structures as well.

In this brief survey of the possible relationships between operators and states we have tacitly considered the eidetic operator as being different from the general state transformation operator. Actually, there seems to be no particularly well-founded reason for this assumption, besides the fact that the eidetic capacities are peculiar to man as the last-developed properties of the human brain, whereas all other properties, are in essence (and in many of the manifestations) little different from those in animals. As we shall see, it may not be possible in every instance to draw a precise line separating eidetic from noneidetic (somatic, autonomic) brain structures, and this could support the hypothesis of the existence of a unique operator for all brain state transitions. In *PTN* the operator of general state transformations was assumed to be comparable in its function (thus essentially with comparable general properties) to a metric tensor. i.e., to the general tensor which induces changes of reference framework, such as the brain states are assumed to be. Through its own components —the metric coefficients— the metric tensor defines the unit of measurement of the coordinates of a Gaussian space (represented by all the parameters of the nervous functions). Since the neural parameters are not all equally affected in every situation or state, the same can be said of the coefficients representing the units of measurement.

Accounting for nonhomogeneity in space, the general theory of relativity introduced the concept of "local" metrics and of the related concept of local deformation of the space time. With this, we can picture two possible situations for a unified operator of brain state transformations, including the transformation of eidetic states. a) The so-called brain diffuse projecting subsystem (with its many components) operates as a metric tensor, thus defining the state of each subsystem, eidetic structures included. b) The complex of the eidetic structures controls the general state of the whole system (probably through the diffuse projecting subsystem). In both cases, the interaction between eidetic and noneidetic processes always allows reciprocal interference in the determination of the general state. Presently, even if certainly no neurophysiologist would subscribe it, the second hypothesis seems to rest on more sound bases. At this stage of the discussion the reasons are not yet clear, but we shall see that it is not at all unreasonable to picture man's brain as fully "dominated" by the intellectual functions —which, after all, represent the characteristics that distinguish man from all the other livings. Every consideration, including also the structural aspects of the problem, seem to lead to the same conclusion.

2 Knowledge and Mind

The arguments in this chapter concern an aspect (a functional property) of mind, perhaps the most representative of its essence, i.e., knowledge. As we shall see, knowledge is not simply a catalogued collection of ideas retained in mind to be utilized when needed; rather, it is made up of the same "substance" as mind —not to be identified with memory, as is commonly done, although in many respects the two are very similar. When one looks back over the philosophical definitions of knowledge from Antiquity to today one realizes that no definition is exempt from subjectivity. In our case as well, the concept of knowledge is heavily loaded subjectively with aspects reminiscent of various philosophical standpoints. Despite the subjectivity of the concept as it is defined, at its basis lies the notion that knowledge forms an integral element in the unique framework in which all the abstract features of mind merge with the material ones into a single "object."

2.1 Essence and Horizon of Knowledge

According to the view outlined in the preceding chapters, presentation of a suitable stimulus leads to a change in the state of the system. This transition is one to a state predisposed to thinking, in the sense that the system enters this suitable framework before the actual recalling or formation of specific ideas. The precondition of this framework transformation for a given function is not peculiar to eidetic processes. Other functions are also consistently preceded by signs (such as EEG and autonomic changes), for instance, voluntary muscle contractions, and such signs may be interpreted as the outward manifestations in the motor subsystem of the framework transformation that is necessary for the voluntarily programmed motor function. (This represents an approximate, but the commonly accepted explanation concerning the central motor system.) Thus, the "stage is set" for the appearance of either movement or ideas. It is against this prepared background that the ideas are formed (or recalled) with their peculiar immediacy and in their entirety, often already assembled in "blocks" of reasoning. The immediacy that is sensed stems from the fact that an idea can be known as such only at the very moment that its elements are simultaneously present —which in the brain circuits occurs at the moment that a feedback is closed, as was suggested in Sect. 1.4. The representational immediacy of ideas or thoughts is facilitated and made more easily understandable by the ideas having been stored in a memory. However, mind has another functional property that may be related to the immediacy of representation as well as to memory, although never identified with the latter —that is, knowledge. We shall dwell on this to clarify a) what the term means; b) how we come to "know;" c) what relationship knowledge has to mind; and d) how knowledge is linked to brain structure.

Let us consider, first of all, the conditions for "knowing." Every known object has its own essential properties, and regarding these an object known "in its irreducibly pure essence" ought to be identical for different knowing individuals. In other words, when the object is known as pure "idea," it is interindividually invariant. However, other factors are likely to introduce peculiar features into the way in which the same object comes to be known by different individuals, while not destroying the invariant character of the idea. The first such factor, one not easily definable, is the way in which each individual knows; this depends on his state of mind (momentary disposition, mood, tendency to distraction) and his general predisposition to know. Other factors, belonging neither to the knowing individual nor to the known object per se, relate to the "surroundings" of the object, both material and immaterial. These factors merely accompany the object and help to place it within a wider "horizon," of which each individual mind gains greater or lesser awareness. These factors represent inter-individual variables and contribute to creating the personal way of knowing an object. In phenomenological philosophy the knowledge of an object "in essence" is called a "reduction," in the sense of reduction to the essential properties. This requires "suspending" all the factors that shroud the essence —those belonging, whether contingently or consistently, to its horizon. This type of knowing can be said to characterize only that in phenomenological procedures; it does not represent the actual way in which an object is commonly said to be known, however. Knowledge is commonly understood as comprising both the object and all its non-essential but normally accompanying attributes, with all the factors and other objects within its horizon, i.e., knowing is strongly affected by the ever-present intra- and interindividual variables.

However, what knowledge *is*, is another problem. It is a metaphysical question which hitherto has received only metaphysical answers, not neurophysiological ones. Indeed, we shall try to provide, if not an exhaustive answer, at least an outline of the theoretical, physiological bases of it. The essence of the knowing process has been identified, by older and by modern philosophers, with the result of a "transformation" by which an object becomes part of the knowing individual, i.e., of his mind-brain. By virtue of this transformation, the object is said to be "appropriated" by the knowing subject. Neurophysiologically, the process must proceed in two steps: first, transformation of the sensorial qualities (if there are any) into neural events; then, transformation of the latter into another, much more complex and articulated, neural phenomenon having the general character of a "recognition," i.e., a first or lower level knowledge. The latter step, however, is such that the object becomes part of the knowing mind, as if "assimilated" to it. The actual physical basis must consist of neural events (nerve impulses), but the result cannot be easily objectivated in terms of the representative nerve impulses —any more than emotions and sensations can be. When the object of knowing is transformed into something like an idea, or a plain idea, it becomes an "integral" part of mind, i.e., the object so transformed actually is the same thing as mind inasmuch as it is made up of the same substance.

We consider now a more precisely defined meaning of "becoming part of" or "assimilation". In this process of assimilation we must maintain the distinction between the invariant part of the object (the essence, strictly, of the object's self, i.e., the idea of the object after a phenomenological reduction) and the variant part (the personal vision of it, peculiar to each individual). As we have seen, the variant part places the object within a wider, but personal, horizon, and makes one person's knowing the object somewhat different from another's knowing the same object. (We might say, perhaps more properly, that the interindividually variant horizon makes us know somewhat different objects.) The process of transformation is therefore more profound than we presumed at the beginning, because the knowledge appropriated by mind includes the complete set of relationships among the various objects of knowledge —together forming, if not a closed unity (i.e., an individual idea), at least a corpus of well-integrated elements. While we may hypothetically consider knowing an isolated object, in a real situation we cannot change or exclude this wider significance given to the object known. Knowledge of an object may only mean knowing it in its more comprehensive framework because, although part of mind, the object does not exist alone in mind. This means that understanding the object entails appreciating the characteristics of its horizon, its collateral properties, and especially its affinities with other objects, i.e., the reciprocal relationships binding all of them. The logical conclusion is that single-object knowledge cannot exist.

Because the interindividual variables condition each mind for understanding the relationships among various elements, ideas, and material and immaterial properties, the resulting knowledge may differ widely from one person to the next. Everyone attains his own integrated, internal vision of the object's characteristics, together with other characteristics not peculiar to it and relationships to objects of other "regions" of knowledge. If the most peculiar property of each mind, and consequently of its way of knowing and of forming knowledge, is the integration —the internal vision of the object within a framework of other objects and relationships— the elements that contribute most to determining individual peculiarity are those belonging to that which we have termed broadly philosophical principles. The almost infinite range of qualitative shades and quantitative grades that each one of us attributes to each of these broadly philosophical factors and to their reciprocal interactions makes the actual framework of each mind wider or narrower, and the resulting knowledge integrated in more or less "broad-minded" manner. In other words, the appropriation, or assimilation, of an object by mind consists in the object becoming part of a "self," i.e., an element suitable also for identifying mind; this self is different from another self, the differences being determined by the characteristics of the whole framework of relationships comprised in mind. But, from this viewpoint, the self, the mind, and the encompassed knowledge are one and the same thing. Consequently, the more integrated and richer in elements the knowledge of a region of objects is, the more integrated and richer becomes the knowledge of other objects subsequently presented to mind, since they are necessarily viewed within a wider horizon —one which they, themselves, in turn will contribute to widening. The eidetic process, i.e., formation of ideas and thought, is the factor continuously at work widening the horizon and therefore the framework of knowledge (read: of mind), i.e., of what are generally called the self's essential qualities and capabilities.

There is a fundamental difference between knowledge of an object after submitting it to phenomenological reduction and knowledge of the same object in its wider horizon. We may be tempted to see one as a better knowledge than the other. However, on the one hand, the former is a "pure" knowledge, concerning the ultimate, irreducible essence of the object, making it an immutable truth, a reference for other knowledge. On the other hand, knowledge which includes the wider horizon is more "complete" and richer, being integrated with that of other objects having their own horizons, which at least partly overlap. Actually, we are dealing here with two different kinds of knowledge. In the first instance, the phenomenologically reduced object is an idea reduced to its ultimate essence, existing outside its relationships to other ideas (it remains one of the unresolved metaphysical questions whether such an idea may have a real existence outside or independent of our minds). Repeating what we have said above, the *Geschlossenheit* of the elements of an idea makes the latter an independent unity, which together with other similarly defined ideas constitutes the substance of our knowledge. In the second instance, the real object of knowledge is neither the single, isolated idea nor the ensemble of the ideas per se but rather the complex of relationships linking the ideas —the sole factor actually attracting the attention— to an ensemble of others. This leads us to the view that the abstract constructions of mind are not closed monads (although through phenomenological reduction they can also be contemplated in their isolated purity) but are connected in greater unities by virtue of their reciprocal, factual, and logical links bridging to the most distant regions.

2.1.1 Knowledge and Reasoning

It can be easily seen that wide-horizon knowledge is a prerequisite for reasoning. It does not constitute reasoning by itself, since the latter is a dynamic process directed to some goal, whereas in that substrate of wide-horizon knowledge the logical paths may be innumerable, each goal dependent (though to different goals) and none actually being followed. Therefore, the reasoning is not determined according to the meaning of the substrate objects alone; only its type, as wide-horizon reasoning, is determined. With this, we do not imply that there may be absolutely no factors that determine thought; some of the latter, for instance, reside in the very nature of the known object. We have noted above that most logicomathematical elements limit logical choices, sometimes permitting in fact none at all —reasoning in this case is predetermined by the very selection of the object, although this extreme is probably never reached. Reasoning about a purely mathematical object may be transformed by the ever-present broadly philosophical elements into a complex type. Since these elements are always present, they may be regarded as constant biases of the eidetic processes —the term bias here not in the sense of inducing "deviation" from a "correct" path. Certainly, this is not the case with the broadly philosophical principles; on the contrary, these contribute to improving the eidetic process about an object by focusing on its different aspects and bringing to light the relationships linking it to other objects, thus making it easier to examine their nature, as though being examined through different color filters. The concept of deviation seems better to fit the interaction of emotional factors with thought; for these a preliminary word of comment is now required.

Both broadly philosophical principles and emotional factors, due to their capability for interacting with thought even within a purely logicomathematical path, are generally considered nonrational from the empirical-positivistic standpoint, and the conclusions reached under their influence as stained by nonrationality. This is not true, of course, and there is no doubt that nonrationality in the line of thought must be excluded in both cases. Emotional factors themselves are integrated into a wide network of logically preprogrammed general reactions, based on a complex of anatomically well-defined control structures and subserving physiological functions; and in this way they are elements of a logical organization. The broadly philosophical and emotional factors are consistently measurable in a way or another through the effects that their mere presence produces, signifying that a rationale is inherent in them, even if not identical in every individual and probably likely to change its unit of measurement in a condition-dependent way. Moreover, the fact that these so-called biases are measurable (albeit by personal yardsticks, or local metrics) means that they may be brought into rational relationships with other elements as well, the precondition for performing the necessary comparisons at the basis of eidetic processes.

While the biases may to some extent affect the line of logicomathematical thought, either directly or by diverting attention, elements of another nature, though bearing the imprint of rationality, do not easily divert the path of firmly established philosophical principles and even less the course of emotional factors (only the outward expression of emotionally determined behavior may be suppressed). The reasons for this asymmetry may be complex, since we are dealing with quite different orders of rational objects. Logicomathematical reasoning is based on "discovering" the logical relationships among the elements and objects of a line of thought (ideas) since these logical relationships are rather rigidly defined by the very nature of the elements themselves. Therefore, in this case mind generally only recognizes the logicality and constructs reasoning upon it. Also here some degree of preceding knowledge is indispensable concerning the framework of laws to which the object belongs. In the case of broadly philosophical principles the situation is altogether different. Far from simply recognizing a logicality of relationships (those in their essence constituting science), mind itself constructs the network of logical relationships in a line of thought, using broadly philosophical elements —products of the same mind and not inherently linked by the same type of logic (necessary and almost independent of mind) that is peculiar to logicomathematical reasoning. We might say that mind adapts itself to predetermined paths in logicomathematical reasoning but constructs its own paths in broadly philosophical reasoning, which thus appears as the more genuine and "spontaneous" product of the eidetic system and the more faithful representative of mind's fundamental capacities.

Thus, mind uses knowledge to build reasoning, but the knowledge consists in the objects which it has appropriated, and which have assumed the same substance as mind. Therefore, reasoning cannot consist of the mind picking up objects somewhere (say, from memory) and linking them to build propositions. The objects are already there, part of mind itself, in what we may call a "phase space" of objects belonging to all regions of knowledge (the nature and properties of this phase space are considered extensively in Chaps. 9, 10). Mind first builds new knowledge from previous knowledge; it uses preknowledge in the most natural way in the continuing knowledge-building procedure, namely in recognizing relationships between the objects. Thus preknowledge is the basis of reasoning, and the latter cannot exist without the former because only objects already known can be introduced into reasoning and logically linked to one another. The task of mind is facilitated when the relationships among objects are clear-cut, as is the case generally in mathematics and scientific; clear-cut relationships, of course, result from transparent logic, i.e., intelligibility of the principles ruling those relationships. Achieving this should represent the principal aim of science

Reasoning is the typical motion of mind, and as such it must rely upon some source of "energy," although most probably not a form that is measurable in physical terms. The secret of mind's energy and its motion must reside in what we have referred to as the relationships among the objects of mind, in the force by which each becomes linked necessarily to others. Thus relationships, links, logic, and principles build a hierarchical set of abstractions within a unique "quality" of abstraction, endowed with the characteristics of both purely metaphysical and matter-bound objects. Accordingly, the whole scheme of reciprocal interactions between operators and ideas, which repeats itself throughout the constant feedback mechanism, should be reconsidered in terms of the possible relationships between the nature of reasoning and the structural organization. For the proposed scheme of eidetic processes to be accepted as a starting hypothesis, the structural organization subserving the eidetic function must provide the capacity for quick management of elements, ideas, and blocks of thought in extremely wide fields and across different regions of knowledge. It must be possible to access the different regions by a "natural" disposition to determine congruence-coherence among the elements of thought, and thus whether they seem to belong to a single natural picture. Introducing two such natural factors into our argument —the disposition and the picture— means that (accepting the theory of *PTN*) the natural disposition to determine congruence-coherence should be part of the brain circuitry in general and not of the eidetic subsystem alone, since all decisions depending on measurements performed by the brain (obviously also those outside awareness) are based on comparisons. At a later stage in our discussion we consider what sort of natural picture would allow the gathering of elements of thought from different and perhaps widely distant regions of knowledge; this is a question requiring the introduction of other concepts (such as the abstraction process, memory, conscience, and the "concept" itself).

2.2 Properties of Mind

We have considered up to this point the relationships, on the one hand, between essential knowledge in a phenomenological sense and wide-horizon knowledge and, on the other, the thought processes. However, we have presented little more than a philosophical definition of what we mean by the appropriation of ideas. We must now ask, in "practical" terms, what we mean by the "mind" of which the objects of knowledge become part, and what it means "to become part" of it. Our goal is to provide a neurophysiological explanation of concepts widely used in philosophy and psychology. The theory as outlined originally in *PTN* must provide a logical place in its framework for terms such as appropriation and transformation in reference to objects of knowledge, from the simplest material objects of the outside world to the most sophisticated ideas of metaphysics. It must also account for the eidetic function. We should be able to come to grips, albeit in a still unforeseable future, with philosophical concepts, as we do now with somatic and autonomic functions, i.e., in terms of neural events. The neural events supporting the former are certainly of much greater complexity than those supporting the latter and are far less easily definable by means of present experimental procedures, but should be no less immediately intelligible. The difference between the two, if there actually is one, should lie in the fact that somatic and autonomic functions are more easily referred to an organization of anatomically defined subsystems, either concentrated within restricted areas or widespread throughout the brain, but easily explored and describable in terms of chemical properties of cells, nerve impulses, and neural circuits.

This is not the case with eidetic processes. These begin, as it were, beyond the limit reached by present neurophysiological methods for exploring the brain. The nature of perception, and to some extent of will, for instance, cannot yet be explained in terms of nerve impulses, although these fundamental aspects of psychic life are still at the borderline of phenomena that may potentially be explained using the more sophisticated methods of modern neurophysiology. They are, ultimately, judgements or conclusions reached through comparative measurements in the course of abstract functions such as recognition, consciousness, or goal determination (see, Chap. 9, why consciousness may also be considered as a function). Although consciousness, recognition, capability for judgement, and other mind functions may be impaired by lesions in various brain areas, so far there is no evidence that they have a "seat" in circumscribed structures. This impossibility in referring eidetic functions to known anatomical structures has diverted the attention of investigators to functions more easily approachable, leaving to psychologists the task of investigating mind. As already suggested, our approach is not based on monitoring the activity of structures but seeks to identify the necessary operations performed by circuits of neurons to penetrate their significance in terms of logic.

We have described the acquisition of knowledge as objects becoming part of the mind; by appropriation or assimilation the object assumes the same nature as mind. This view still consists of figurative expressions, useful for presenting a certain understanding of the results of a still obscure process. To explain what the expression means, however, when we speak of objects "becoming part of mind" we must begin with a notion of mind as something that is sufficiently preformed and stable as to maintain its individuality throughout life; it must be seen as a real object which, though abstract in nature, occupies a real physical space and is thus, in some way "in the brain." As a familiar verse goes:

"Alma in verse, in prose the mind../Sits cock horse on her throne the brain.." (from E.D.Adrian, *The Physical Background of Perception*)

We must yet consider the "what" and the "how" in terms of structure and organization of the brain. Some of the metaphysical aspects of mind are dealt with in the "Appendix" of *PTN*. There it is stated that the "intellect" of man is the sum of the powers of his *anima intellectiva* (the intellective soul) peculiar to humans. It is clear that the intellective powers are those of mind (*alma*, in Latin, means "life giving," and is thus related to *anima*). This may be a mere question of semantics, and one may not wish to define so closely intellect and mind, since the two terms commonly refer to properties of somewhat different natures. We insist, however, that they actually represent properties of a single abstract structure of our brain. If we wish to distinguish them, we can view mind as the "active" side of thet which the intellect is "potentially." This is a consequence of the mind being preformed and stable, as noted above. This property can be attributed only to a "structure;" thoguh conceptually abstract it is a thoroughly matter-bound structure for it is also organized functionally in brain matter.

We still lack a "what," an indivisible unity graspable in its entirety. Although now endowed with such an "effective" capability, we must expect to find this implemented in the brain's structural organization. Even if this is not graspable in its entirety, we must be able to observe and possibly to measure it through its products, i.e., logicomathematical and complex thoughts, memory, consciousness, etc.—all the properties of man's brain beyond the purely somatic and autonomic sphere. Mind therefore refers to the complex structural organization embodying the intellectual powers in practical terms (while "powers of *anima intellectiva*" is merely a metaphysical concept). Through its innumerable connections within the brain matter, mind is omnipresent and controls the whole-brain processes; we can thus say that mind is represented in the whole brain. (This conclusion closely agrees with the principle, as stated in *PTN*, that there can be no clear-cut boundaries between various functions of brain and no strict localization of the related processes.)

While mind is in this way bound to matter and enjoys the same degree of stability as matter, it retains a great capability for changing qualitatively because its relationships to matter are principally in terms of organization, and the latter is, by definition, based upon changing patterns of relationships in functional terms. Nevertheless, these changes do not entail a loss of power since the modification of mind is the continuous adaptation of its structure by conforming the activities of the different subsystems to the requirements of maturation, learning, knowing, etc. These processes do not undo the results of previous maturation, learning, or knowing; the changes which they introduce are merely in the form of improving function. In other words, there are no substitutions or mutational changes but a continuous perfecting of existing powers. The system seems capable only of improvement; it can never deteriorate spontaneously. Whenever a deterioration does occur, this is due to loss of the physical substrate and thus of neural circuits for organization. The system's being able only to improve is implicit in the view, elaborated above, that an eidetic operator can be modified by continuous interaction with the outcome of the operations, bringing about only a progressively more sophisticated performance of the whole system. Mind therefore possesses some enduring properties which may nevertheless be subjected to change, although not over short periods of time.

Since mind's properties constitute the peculiar characteristics of man, we should be able to find traces of these enduring qualities in some enduring aspect of the brain's structural organization. The "Appendix" of *PTN* contains a number of observations on this in relation to the continuous activity in brain circuits. It is noted there that man's peculiar capability for judging external events as well as his own thoughts cannot have been learned from animals or derived from inanimate matter. Therefore, it must somehow have been imprinted in his matter by learning either alone or together with a genetically inherited "composite" made up of structure and particularly stable patterns of functional organization peculiar to human nature. With these basic properties of structural organization man's brain judges the world with its own peculiarly human standards of comparison. The latter, however, must not be confused either with memory patterns, which man uses to compare that which comes from without, or with his own thoughts; it is obviously impossible to think of a memory-predetermined pattern for each of the myriad different situations that the material and intellectual life of man may encounter. The inherited structural organization can only embody the complex of bonds constituting the principles of the processes leading to judgement, ultimately representing the standards with which man's brain probes the external as well as its own internal events. Nothing, either of the external or of the internal, escapes mind's judgement. This is true for every object and every contingency in life ("Deus constituit hominem et reliquit eum in manu consilii sui," to quote from *Ecclesiastes* 15 <14>; that is, God created Man and left him in the power of his own judgement). If there is anything in the structural organization of brain that exhibits enduring characteristics (while still capable of long-term adaptation), anything that serves as standard for comparison through the fleeting traces of multifarious events, this can only be some stable (and thus possibly recognizable) components of the continuous activity of brain circuits. The stable patterns of this activity must present a basic field of "deformation, " i.e., the expression of a "strain" as represented by a tensor, the actual impression in the brain structural organization left by the field of forces which maintains that basic distribution pattern of activity (with which every encountered event must be confronted, see *PTN*).

The existence of a strain tensor (and the related stress tensor) is not a mere presumption, since it is a necessary, unavoidable corollary of the continuum character of ongoing activity due to the brain's structural organization. The presumption is, rather, that it may exhibit basic patterns peculiar to man and embodying the reference standards that man uses for comparing the qualities of the events with the peculiar quality of his own nature. To build its own body of knowledge requires of mind the capacity to recognize the object suitable for assimilation —which is its actual purpose and thus the purpose of this comparison. The latter also is a far-reaching presumption. The set of basic principles for comparison that is represented in the presumed strain tensors should allow man to cope not only with reasonings required in logicomathematical regions of the eidetic function but also those in philosophical, ethical, social, and religious ones, through which man can meet all the possible needs of his intellectual life. One might object to this proposition on the grounds that it implies an inherited (genetically determined) and hence predetermined property of man's mind. However, we should consider that mind's qualities have made man the being that he is. This justifies hypotheses about a possible genetic coding of the basic principles for comparing criteria, or measuring standards, of the inner with those of the outer world. What is presented to mind for appropriation is already the result of a series of transformations; we may consider these as a kind of previous "formatting" of the objects (such as occurs through the sense organs, central analyzing structures, and perhaps "external" language, see Sects. 10.6-10.6.2), so that the comparison may take place only at some advanced stage of abstraction in the eidetic process. From this point of view it is easier to understand how, by continuously exercising the intellectual functions, the characteristics of the basic field of deformation expressed in patterns of ongoing activity becomes progressively more suited to them. The neurological procedure of comparison, as the basis of eidetic processes, also involves the continuous interaction of the eidetic object with the basic strain field, an operation that tends in the long run to shape the basic field progressively better to fit the object. On the other hand, the hypothesis of genetically predetermined patterns of strains shaped exclusively by interactions with externally supplied objects cannot account for the extreme variety among concepts related to immaterial objects of mind's concern; neither can it provide a suitable basis for their explanation or for the intrinsic freedom of man's mind from conditioning of the surrounding world (see Sect. 5.4). The extent of such freedom is discussed in the "Appendix" of *PTN*; consistent with the conclusions reached there, we conclude that the characteristic principles of comparison must be part of man's mind, independent of specific, acquired impressions from without —although they may be modified in the course of life through continuous interaction with eidetic processes. This interpretation is consistent, on the one hand, with the long period required for man's intellectual maturation and, on the other, with the presumed possibility of transformation of the eidetic operator. For this, no memory would be required, in the sense of a physically stable storage of either events or eidetic objects, only the capability, less for retaining than for changing the properties of the basic field of strains (we shall return to this point below).

We have interpreted the metaphysical concept of appropriation as the original raw objects (or their transformation into ideas) being known through the "internal vision" of mind, which also endows them with its own personal qualities. The actual neural operation involved here still cannot be defined (we shall not even attempt to tackle the problem). The essence of "knowledge" also remains imperfectly defined, although as a concept it now appears more intelligible. We may use the more descriptive term "integrated vision" to denote the mind comprehending simultaneously the object and its accompanying properties —the halo of the surrounding objects in the wider horizon, the relationships with the objects of other regions. As was suggested above, the objects do not need to be memorized in their "appropriated" form, as stable constituents of a physically conceived mind. The process of transformation, which "introduces" the object into mind, may also be conceived as occurring every time the object is presented. Only the means of appropriation must exhibit some degree of stability, although a certain variability in the result is unavoidable due to the ongoing interaction of the object with mind's present knowledge. This is probably the reason of the fact that hardly a thought or a memory is exactly the same when resumed at different times since the eidetic processes, through which thought and memory are created, can never undergo exactly the same development. (This will be discussed in Chap. 7 in relation to memory.)

2.3 Geometry of Mind: Space and Type of Thought

We concern ourselves now with some further aspects of the question of brain "space" in relation to thought processes. The conclusions reached in *PTN* and the preliminary remarks in the present monograph support the hypothesis that the very act of thinking is by itself sufficient condition for, and actual cause of, a change in the "geometry" of cerebral "space." We should recall that the latter is a space of functions of neural parameters, and that what we may call motion within this space is a change of state of the system (the state being the resulting function of all the variables involved). If we liken the mind's attentive engagement to an accelerated motion, the thought processes influence brain space time, as every accelerated motion does, according to the principles of general relativity. In the accelerated state of motion, time interval judgements performed by test subjects have been reported to reveal a deviation of "personal" time from clock time; however, the relationship of personal time to "proper" time (the *Eigenzeit*) of the cerebral system is still unknown because as yet no attempt has been made to determine the latter.

The question is whether purely logicomathematical thought and complex thought (possibly with accompanying emotional factors) differ in terms of properties of brain space. There is no doubt that both types of reasoning may grasp and hold the attention, so that these should not at first sight differ in their respective capability for accelerating brain motion, particularly when we consider that these two types rarely occur as strictly separate and in pure form. However, to be more precise, the problem is not exactly whether the two types of reasoning influence the geometry of brain space differently, but whether each represents a distinctive geometry. When talking of brain space and brain motion, we should not forget that, as in physics, our reasoning makes sense only insofar as we also adopt the framework of relativity, according to which space does not resemble a container filled with matter moving along space and time coordinates, but space itself *is* all the matter and energy present, as well as time. Thus, we cannot represent the thought processes of either type as motions within cerebral space but must ask what kind of space these motions constitute (for which we may define space, time, and energy parameters) —in other words, what is their geometry. Considering the two types of reasoning from the point of view of their constituting element-ideas, substantial and fundamental differences stand clearly out. Logicomathematical thought may be said to be pure inasfar as it consists of chains of ideas that are largely predetermined. This means that the intrinsic logic —the logic holding these ideas together— makes up a chained system of objects almost perfectly predictable as to the possible course. The predictability of this motion in turn rests on the fact that, given some sets of elements as antecedents, one can always reach conclusions following an inescapable logical sequence, defined by the precise, predetermined relationships among those elements, i.e., following precise rules or laws. The particular nature of the elements of logicomathematical reasoning and the fact that their characteristics are based on uniquely defined, precise relationships among the elements makes this reasoning invariant and in some sense also independent of time, i.e., it is largely invariant both intra- and interindividually, and is therefore not individually specific and is to a large extent immutable. This implies that logicomathematical reasoning generally affords a symmetry since, if the precise, predetermined relationships among the elements lead to prediction of a unique result from a given set of antecedents, it must obviously also allow retrodiction of the antecedent set from the result. This symmetrical property characterizes only pure mathematical thought, and it vanishes progressively when the elements are no longer strictly logicomathematically linked, or when elements of a different nature are introduced —in short, when no unequivocal antecedents and results can be derived, as is general case in other types of thought, including generally scientific thought.

The similarity of the laws of logicomathematical thought with other laws, for instance, Newton's law of motion, is clear here. The Newtonian law of motion allows an almost absolute calculability of future as well as past motion, and by stating only the relationships between the factors without reference to a fixed time it is also time symmetrical. The typical framework reference for Newtonian motion is Euclidean, the space providing the reference for all inertial, or time-independent motion. Is the latter perhaps also the reference space for purely logicomathematical reasoning? In other words, are the characteristics of mind's space, when mind is concerned with purely logicomathematical elements, also Euclidean, i.e., is the motion equivalent to these eidetic processes indeed of the time-independent, inertial, constant-velocity, type? It is clear that this type of motion, inasmuch as it exhibits a constancy in time, is the kind of steady-state motion sufficiently resembling natural motion in the sense of statistical mechanics, as well as a perpetual eidetic motion based on the steady-state, ongoing activity of the eidetic structures. Here we observe two aspects of the same abstract structure: logicomathematical thought as constant-velocity motion, i.e., the symmetrical formulation characteristic of this thought, and the constancy of results and content (since the thought's elements are in strictly predetermined, almost rigidly fixed relationship to each other), entail an almost absolute predictability.

We conclude from this that logicomathematical reasoning, possibly resembling inertial motion, gives rise to a smoothly proceeding, almost necessary development. Such reasoning may nevertheless be disturbed by "forces" apparently accelerating the eidetic process by introducing into it other elements, with a resulting motion no longer representable through a simple time- and space-symmetrical formulation, a thought without exactly predetermined relationships among elements, and having a substantially reduced degree of predictability. However, these forces cannot change the fixed relationships between the logicomathematical elements inherent in the latter and constituting the basis of thought's predetermination and consequent predictability. Rather, the "accelerating" new elements, by interfering with the judgements of congruence-coherence among the ideas of thought, may diminish the capability for clearly and unequivocally recognizing the mathematical predictability. In other words, the accelerating forces appear to divert the logicomathematical reasoning from its predetermined path but actually affect the capability for recognizing congruence-coherence.

As a corollary, for the system providing the judgement on congruence-coherence the Euclidean framework no longer represents a sufficiently reliable reference for the eidetic processes, and unless transformation to another, more suitable reference system is provided (according to general relativity principles) —i.e., a new local metrics that may "straighten" the distorted measurements of congruence-coherence by restoring covariance of the measures— the reasoning is bound to suffer considerably in the logicality of development. In line with the interpretative theory advanced in *PTN*, cerebral forces may be able to "curve" brain space time, as physical forces (e.g., gravitation) do with the physical space time of our universe. Thus, on a relativistic basis, we may picture the change in the state of motion as an automatic adjustment of metrics, the transformation of reference space time coordinates, from the four-dimensional, orthogonal coordinates for inertial motion, to a Riemannian space, curved coordinates, no longer Euclidean reference system. (Actually, for brain space a particular type of reference is assumed, the Gaussian coordinates system, exhibiting either Euclidean or Riemannian space characteristics at different points, thus allowing transformations of the reference, still retaining the possibility of invariant measurements. We shall not resume here a discussion of the theory elaborated in *PTN*; suffice it to say that the characteristics —Euclidean, Riemannian— in different regions of brain space are taken as a function of the forces acting locally on the elements of the space and thereby introducing changes in local metrics into the system of coordinates —the variables involved in the brain function.)

It seems all too natural to assume that the forces which disturb the smooth development of logicomathematical reasoning and make it less predictable or altogether unpredictable are elements of broadly philosophical reasoning (strictly philosophical, ethical, religious, social, etc.) or features of emotionality. These are not characterized by rigidly fixed reciprocal relationships; quite the contrary, they offer at each step of thought many possible alternatives for choice and therefore for comparison (measurements of congruence-coherence). Thus, at each step a judgement is required, but for the latter the weights of the respective measures may be changed according to as yet undefined rules. However, the changes in the weights associated with each measure should be in direct relation to, or directly represented by, the changes in coefficients of a metric tensor, that bringing about the change in units of measure in the regions of space subjected to the forces. All of these "disturbing" elements possess a double aspect. First, as just noted, by directly or indirectly evoking the change in a metric tensor they transform the general reference system for brain functions; thus, at least theoretically, they may interfere with the smooth development of logicomathematical thinking. Secondly, they also enter directly into the mechanism of judgement and of decision as elements on which both judgement and decision are based, together with the logicomathematical ones. The consequent change of Euclidean to Riemannian space (from orthogonal to differently curved coordinates) results in a reference system no longer suited to purely logicomathematical reasoning but now better adapted to the broadly philosophical one.

These arguments may appear to lead to conclusions somewhat in contrast to those made elsewhere (see *PTN*). The reason for assuming a change of reference system is the experimental evidence that the measurements of subjective time interval consistently exhibit deviations from the intervals measured by a clock —this entailing a typical general relativity situation of accelerated motion— when the test subjects are engaged in sufficiently demanding mental tasks, regardless of their actual content. This therefore discriminates only between mentally active and mentally idle states. The findings of a slowed subjective clock in these conditions introduces a disagreement with the present hypothesis that logicomathematical reasoning, while certainly putting heavy demands upon our mental powers, at the same time is represented by a Euclidean reference system. In fact, if the relationships between the elements of logicomathematical reasoning —ideas and sets of logically connected ideas— are representable by symmetrical laws, similar in structure to Newton's law for constant-velocity inertial motion, an accelerated motion is automatically excluded. On the other hand, since broadly philosophical reasoning is also rationally based, the elements which constitute it must also be linked by relationships having the form of physical laws. This observation, however, should not obscure the fact that there is still substantial difference between logicomathematical and broadly philosophical ideas, inasmuch as the latter are never as strictly and necessarily linked to one another in the reasoning process as the former are. Therefore, the difference between the two types of reasoning may in part lie in the general form of their formulation —such as, in philosophical thought, asymmetry and nonlinearity of relationships among the elements— but in part also in the intrinsic state dependency of the units of measurement used in the fundamental operation of comparison.

This problem must be examined more closely. In *PTN*, the relativistic deviation of subjective (personal) time from objective (clock) time is seen as being related to the nature of the mental task; if this consists of uninteresting items, or if no task at all was imposed, no deviation was seen. Thus it is less the task itself than the "reaction" or simply the "feeling" of the subject toward the imposed task (or the general situation) that makes the difference and therefore causes the change of reference system revealed by the time deviation. The discriminating factor, in other words, is the subject's attitude toward the task. This interpretation accounts for inter-individual variability associated with factors such as mental disposition and capacities, knowledge, training, mood (and, of course, the ever-present genetically inherited characteristics) and explains the fundamental unpredictability of behavior, even in the presence of rather simple situations. Seeing the discriminating factor not to be the task but the subjective reaction to the task, we then find emotionality among the most effective reaction-determining factors. We have described "feeling" as an ensemble of reciprocally interacting mental and autonomic elements, often undefined although capable of overt expression and presenting an infinity of shades and different degrees of rationalization. (The role of emotionality in altering the course of eidetic function is examined below, Sect. 5.5.)

According to this interpretation, two different situations may occur in the course of purely logicomathematical reasoning. Although certainly made up of a chain of logically inescapable deductions of ideas, this type of reasoning does not always present the same possibility of immediate recognition of succesive steps due to an intrinsic difficulty in following progressively increased levels of involved abstractions. This gives rise to the mental "stimulus;" mind is challenged to meet the demand for full attention and endurance. A heavily demanding task and the readiness and willingness of mind to meet it determine the transformation of reference system, i.e., the appearance of space characteristics suitable for this mentally engaged state. In perhaps no other instance more than here is the term "attraction" appropriate. The "attractiveness" of a task results from the judgement performed by the subject, who feels almost compelled by a "force" inherent in the nature of the problem. Attraction is the physical force reciprocally exerted by two bodies, and "attractionlike" is the force reciprocally exerted by the abstract structures involved, those of the problem, on the one hand, and those of mind, on the other. Only a force, no matter how we may objectivize it, can be the determining factor in the relativistic transformation of reference system, in the physical, inanimate world and in our living brain. The conclusion is not reached on analogical grounds but on what we believe to be the real conditions of brain, where real, in essence neural forces are active. (We consider below the more specific "eidetic" forces, Chap. 4.)

One might suppose the most attracted and interested in logicomathematical reasoning to be the mathematician himself. This is true, but only inasfar as the mathematician feels mentally stimulated to solve a problem. However, the average mathematician confronted, for instance, with a simple logicomathematical concept, such as first time derivative, certainly does not feel the same type or degree of stimulus attraction and mental engagement as might be expected of an amateur student or a schoolboy. For the professional mathematician the concept has become a mere instrument in his everyday language, not requiring investigation to be understood and correctly used; therefore, one may hardly expect his brain motion to be accelerated and his personal time slowed down (no more than one could expect his watch to be slowed down).

This possibility of a diphasic mental attitude is not likely to appear when dealing with broadly philosophical reasoning, either pure or associated with logico-mathematical elements. In general, thought that is philosophical, ethical, etc., although perfectly rational and logically structured, is for a number of reasons bound to engage the mental processes strongly. The outcome of this type of reasoning is far from easily predictable, a fact that distinguishes it from purely logicomathematical reasoning. Previous knowledge may even be of little or no help, in contrast to mathematics and generally to all natural sciences. The ideas begin with poorly defined "images" of still hazy concepts and even less precisely defined elements; the thought formed upon these ideas must then undergo successive steps of refinement. The manifold meanings of philosophical, ethical, and religious elements may be objected to, questioned, and debated, with the result of presenting thought with a variety of dichotomizing paths, making every step a problem of its own. The main hindrances in this type of reasoning consist in the not universally accepted terminology and in the absence of sure standards to which to refer. Furthermore, each element in the reasoning generally presents greater difficulties of abstraction than those in logicomathematical thought, since they represent concepts and ideas not related to the physical world and therefore lacking direct experience and empirical evidence.

Nevertheless, philosophical, ethical, religious, and social elements of thought, while apparently belonging to separate cathegories of their own, actually cannot be separated in principle from logicomathematical and broadly scientific ones. This is not an arbitrary, personal presumption. Knowledge of the concept-ideas in both fields, as extensive as can be afforded, will help us eventually to realize that all the elements of eidetic function are closely interrelated, provided mind can attain the suitable level of abstractness of reasoning. Of course, this requires mind to be thoroughly trained in developing eidetic activity in the most variate fields (see below). For these reasons it is likely —and expectable— that cerebral eidetic activity should exhibit all the general characteristics of accelerated motion, therefore typically moving in a correspondingly non-Euclidean, Riemannian space. However, as just noted, the very fact that mind is involved with either logicomathematical or broadly philosophical problems, thus with the respective elements and types of reasoning, does not by itself determine the type of motion, because matching is needed with the corresponding attitude on the part of the subject, the mind's disposition to attraction. This means that to a subject, who is either culturally untrained or is trained mainly if not exclusively for low-level abstractions differing little from those in the simple experimental paradigms of animal tests, the logicomathematical but particularly the broadly philosophical problems will appear dull and unattractive. Consequently, the subject's eidetic processes are not engaged at all, and his brain activity may remain unaffected, so that his Euclidean brain space would suffer no strains.

Obviously, this does not imply, however, that the mind of a mentally untrained subject never undergoes accelerated motion. The culturally untrained mind reacts strongly to the objects of its normal training, such as everyday experiences, stimuli and situations toward which he feels attracted or is compelled to react (such as reading comics, watching television, driving a car or a motorbyke, making love, eating, etc.). This individual may be able to connect different experiences and reach some suitable level of integrated knowledge of the physical world, which means that he can abstract from experience and build abstract structures. If the determining factor for inducing brain space deformation is interest and therefore the "attraction," we may easily predict that the brain space of culturally untrained subjects will undergo deformation as frequently as that of cultivated ones.

Part II

Problems of Thought

3 Modes of Thought

It would not make much sense to treat the problems of thought separately from those of mind, but it will become apparent that the human mind has developed thought along two principal lines: analytical and theoretical. The abstractions suggested by the outer world and those originating autonomously in the mind are thus dealt with by two modes of reasoning, the elements treated not being mode specific. The mode chosen is determined by mind's purpose. The differentiation into two modes reflects the dual nature of mind's interaction with the outer and inner worlds: on the one hand, the need to understand the mechanisms of causality of specific events and, on the other, the need to understand the rules which govern all events. This is not a personal interpretation of mind's operations; a close examinationof the paths of reasoning reveals the existence of these two modes, distinguishable in terms of the logic at their basis —which is thus the essential difference between them. Difference in logic, for its part, makes attaining the level of covering laws by starting from the analysis of the mechanisms of specific events logical impossibility; however starting from the level of the covering laws should always be logically feasible to reach every possible mechanism governing the development of events. The latter proposition contradicts the separation of logics, but it clearly suggests the practical logical impossibility of reaching knowledge of *every* mechanism of causal development because to do so would require knowledge of every possible antecedent for each event (up to the "primary cause"). This being beyond the reach of the human mind, the hiatus between logics persists as the impossibility of following with an unfailing logic the retrogressive path from result to antecedent (as was stated explicitly by Einstein). In fact, the existence of two different logics is in itself preposterous, since contradicting logics cannot exist in our space-time (whereas, somewhat different logics may exist in different spaces), and it must therefore be understood that the two apparently different logical paths must merge at some point. The theorizing mode is of particular importance in this monograph for two reasons. First, it is the purpose here to represent a theoretical (and as far as possible an operational) approach to the eidetic process from the point of view of neurophysiology. Secondly, as will become apparent below, the theorizing mode represents "truly" eidetic process, the "real" mode of thinking, through which mind makes use of all its capacities.

3.1 Analysis and Theory: Drive to Theorizing Thought

Here we confront a central problem in the investigation of mental functions. Thought is itself an activity that requires a great deal of introspection together with the capability for analyzing it. We shall see that, from now on, the problems will deal with matter not yet touched by physiological investigations. Their solution, or the proposals put forward for their solution, cannot fail to strongly influence our personal attitude as scientists toward research in general, including philosophy, ethics, and religion and to have immediate impact on our behavior. The fundamental problem is whether the structural organization of the brain, to the extent that we know it, in some way conditions the development of thought. This should not be seen simply in terms of setting limits to our reasoning capacities (which is almost self-evident) but as possibly providing "preferential" paths leading largely to predetermined solutions, though not easily foreseeable beforehand. We are not dealing here with the "types" of thought as described in Chapt. 2, i.e., logicomathematical and broadly philosophical; these are differentiated according to their content —their objects— and appear in pure form only at the extremes of a wide range of possible mixed-type reasonings. Instead, we are concerned here with the "modes" through which thought develops, independently of its object. Actually, we have already suggested that the modes of thought development may be independent of their content.

Let us ask first whether the mode of thought is differentiated according to purpose, i.e., to the goal of the reasoning. This requires, on the one hand, describe various purposes of reasoning, not dependent upon the object, and, on the other, that we recognize different structures of thought with reference to different purposes. In the very general case, we may assign to thought only a few fundamental characteristics, not entirely independent of one other, i.e., the development of "what," "why," "how," and "finality" of the object. The questions of what and finality, on the one hand, and those of why and how, on the other, are variously interrelated in the course of thought, depending partly on the nature of the object and therefore not possibly affording each a peculiar mode of thought. However, upon closer investigation differences in fundamental nature stand out. Asking questions about "what" aims at the "essence" of the object while speculating about "finality" is both directed and limited by the depth of knowledge as to "essence." "Why" is much less inherent in "essence" per se but is particularly and indissolubly bound with "how" the object develops in a chain of causes and effects towards the ultimate event. Thus, on the one hand, we have reasoning directed toward the "essence," i.e., the immutable, and on the other, reasoning about a "development," i.e., the continuously changing. We must take heed of these considerations in the following discussions.

In the "Appendix" of *PTN*, we consider only the "scientific" modes of reasoning: the analytical-experimental and the spontaneous or theorizing. These two modes of reasoning support the mind's excursion through the whole field of possible knowledge, leaving untouched only border areas of thought not based on strictly logical connections between elements, such as mystic and poetry. In the latter two, deeply suffused with emotionality one may also find the logic linking ideas and sentences —not with a formal but a purely intrinsic coherence. Eidetic processes are not altogether independent of emotional elements and may actually be profoundly influenced by them. Nevertheless the eidetic function may proceed almost undisturbed when, given suitable training, the mind succeeds in attaining a "suspension" of the not strictly logical and scientific elements, thus excluding emotionality from thought development. The two procedures of reasoning about general problems, including all scientific and broadly philosophical questions, i.e., the analytical-causal and the spontaneous-theorizing, seem to follow the same path, although aimed at opposed, if not always clearly precise goals. They also subserve different and differently appreciated needs. On the one hand, analytical-causal thought aims at discovering the ever-finer mechanisms of material as well as nonmaterial processes concerned with both biological and spiritual fields. On the other hand, spontaneous-theorizing thought aims at formulating general principles and a general framework of laws, i.e., a general theory covering the contingent observations.

In the "Appendix" of *PTN*, we present reasons why neither analytical nor theorizing thoughts can be dispensed with in seeking most complete knowledge of man's brain that is possible. By following the uninterrupted chain of causes and effects, of antecedents and results, studies distance themselves ever farther from the principles which regulate the phenomena, i.e., ever farther from the essence of their origin. Therefore, mind must revert to formulating general laws which allow one to predict the development of all phenomena. This involves an insurmountable obstacle, however. The mode of thought that guides the analysis of cause-and-effect mechanisms is not suited to retracing the path to the origin of the chain, i.e., for developing the more general framework of principles. What prevents this, is a logical fault, an hiatus in the logical continuity of thought which, seeking to use the observation of contingent events (or experimental facts) to explain the causes, can only fall into a tautology, a rephrasing of the observational or experimental data. The reason is that the cause of a specific event is itself a *specific* cause, one for that event alone, sometimes among a great many slightly different causes, and cannot be taken into account in the case of events even only slightly different. On the other hand, no thought and no natural event is so perfectly well understood that it can be absolutely and uniquely determined for each step what the specific antecedent was that produced it. In fact, the certainty of physical cause-and-effect relationships is questioned when the number of variables and the degrees of freedom increase (and also breaks down at quantum levels). The same consequentiality of mathematical reasoning fails when logic is pushed to the point of probing the truth of its own basic postulates and the absolute certainty of the conclusions. Purely consequential, analytical logic is thus fundamentally ineffective when applied in the retrogressive direction to the chain of causes and effects, and mind must find another way of reasoning, that is, must follow the logic along another path. Thus, we focus here particularly on the spontaneous (i.e., not consequent on anything), direct, primarily theorizing thought which, not making recourse to analytical procedures, may provide the sought-for corpus of theoretical principles for making predictions to be further submitted to verification on the basis of analytical-causal tests.

There are further reasons for being concerned primarily with theorizing thought. In the analyzing procedure, mind's attention is centered on the "object" resulting from an experiment (either mental or physical), the unknown x at the end of a chain of causes and effects up to that point. The simplest paradigm is: if facts A and B are given, then fact C follows. The search for a reality based on this type of relationship may lead only to the discovery of an infinity of such factual realities, since to the mind's eye no objects are offered except those presented to the physical eyes by naturally occurring or artificially created contingent situations. The danger is that by an unacceptable twist of logic the investigator is led to believe that the essence of his problem consists in the myriad elementary essences discovered as realities, and that no other reality exists than that which his eyes and his instruments can detect. This fallacious attitude usually goes unnoticed because today's powerful means of investigation allow one to continue discovering ever more truths, ever more scientifically interesting facts, down to the inner core of matter, following the self-sufficient, undeniable, matter-bound logic of Nature. According to the most unquestioned tenet of empirical-positivistic science, one must leave it to nature to guide the investigation, since only the intrinsic, unfailing logic of nature can suggest the right path to follow; however, this may be true only for observations in the course of an experiment. (Paradoxically, the same error may be caused in the field of the broadly philosophical sciences —philosophy, ethics, religion, politics, etc.— by the accumulated wealth of critical studies entering a maze of analytical, subjective interpretations which easily lead reasoning astray and toward the inessential.)

The modern philosopher Fabro observed "L'homme en réalité s'est emprisonné dans la loi de la conséquence"; that is, man has imprisoned himself within the law of consequentiality. If we want to break this chain —which can only lead (or push?) us ever more deeply into the myriad contingent details and make us confuse mere attributes with essence and then essence with *esse* itself— we need only to claim the natural independence of our mind and address it to the other end of the line of thought: at the origins, at the beginning of the causes, where "what" and "why," "finality" and "how" are present all in a single framework, in a "theory." However, since there is a gap that logic alone cannot surmount, thought must seek to independently and spontaneously build that framework; this must then be tested further by experience to satisfy the scientist as to its suitability as a faithful representative of what." "why," "how," and "finality." Man's mind can do this through its natural speculative spontaneity (*spontanéité spéculative*, to quote the philosopher and theologian Dubarle) —its peculiar, intrinsic capability for freely, directly speculating about the laws of our physical and spiritual universe, up to the origins, the *esse*, the "act" at the origin of the being. There is no doubt that man can do this. He is endowed with an apparently natural capability and natural disposition for freely speculating about the order of things, a property certainly peculiar to humans, based on the intellectual properties unique to their minds. It is not peculiar to scientists but inheres in all ingenuous minds. Surely, it requires rigorous training, besides vast knowledge, to produce theories that withstand the test of criticism and experience, no matter what the object of reasoning is. It is the mode of reasoning itself, i.e., theorizing, which requires training and knowledge. Our books and scientific journals present us with plenty of so-called theories which are, however, no more than working hypotheses, founded on some experimental data, and seeking principally to explain them despite myriads others data. A true theory is the one that can explain the whole set of possible experimental results. However, without rigorous training and vast knowledge, free speculation may also succumb to dangerous pitfalls, such as when one shuns preexisting knowledge and one judges it unnecessary for the development of an overall teory.

3.2 Analytical and Theoretical Procedures

We have not yet touched the core of our problem: a) whether the structure of thought remains essentially always the same or depends on its purpose (analysis versus theory) and b) the part played in the structure of thought by brain's structural organization. In principle, we consider analytical and theoretical thought to be different regarding their goals, although a great deal of confusion still exists as a result of the terms not always being clearly defined. Particularly the term analysis is used with widely varying meanings. If strictly analytical procedures produced results having the nature of general laws, we could presume an identity between analytical and theoretical thought structures —and that theory and analysis are performed by the same neural structural organization, albeit using different strategies (nevertheless, it would be difficult to understand how all this might occur).

The Greek root of the word analysis means "to break up," to "fragment." Thus, using analysis to solve a problem entails decomposing the object into its component parts and, as was apparent in the preceding section, discovering how the parts are related so as to constitute a unified whole. However, particularly in the realm of experimental biology, a true, strictly rationally defined goal is not always present, or at least is not sufficiently clear in the mind of the experimenter, and this in turn diminishes the efficiency of the method. For want of a clear, rationally defined purpose, the original starting point can be formulated only generally, and to be of any use it must be further divided into subquestions until one obtains a sufficiently defined and circumscribed argument to be submitted to analysis. Therefore, generally the procedure becomes on of descending to particulars, very similar to characterizing an ensemble through the description of its parts, the procedure nowadays followed by the great majority of neuroscientists. The latter frequently define the experimental procedure as a search of the causes or of the roots of the events. A search of the causes means to seek that which precedes, but if we follow the course of thought in these cases, we realize that the object of the search is actually fragmented into a set of hypotheses.

The word hypothesis also derives from a Greek word, meaning "something put under" and thus a "basis" or "principle." The modern concept is that of a schematic picture or outline of the essentials a system, i.e., a proposed description of properties needing experimental confirmation. However, no true hypothesis can be formulated with a thought procedure that descends to particulars (as analysis does), since to build an hypothesis requires the capability to look backwards over the path of thought to a "whole" that comprises the particulars. The path of an analytical procedure, if followed backwards, leads only to justifying the very fact of that backward motion, i.e., a mere rephrasing of it, hardly different from a tautological procedure. This means that analytical thought is altogether unsuited for retracing the origins, and must be replaced by that type of eidetic process which alone can produce a true hypotheses, i.e., a theorizing motion. The more the starting point is submerged in the details of a general problem, the farther the character of the hypothesis tends to be from that of a possible framework for a system. This takes on the quality of what we may call the "probable" cause of the observed fact, generally called a "working hypothesis" (we may say, better, a "working model"), i.e., again an assumption valid only within the restricted range of a few particular phenomena, or even for only one of them. With it, the scientist may proceed only stepwise, usually indicating as starting point a single proposition, containing the supposed truth to be tested by experience (see the next section), and needing to be answered simply with "yes" or "no."

Analytical reasoning is different when one is dealing with the physical events of inanimate nature. Physics is deeply rooted in mathematics, on the one hand, and in the physical world, on the other, but the physicist does not initially use the empirical approach to analyze directly the causality chain of an event as is produced in nature (or does so only in particular cases). Generally the empirical procedure is the last step and aims at testing a proposition. Rather, the physicist approaches an event by "reproducing" it in the laboratory or on paper only after careful definition of the "initial conditions" (or the initial state). Moreover, the "object" is not investigated per se but as a representative realizing its originating principle. Therefore, the physicist also descends to particulars along the chain of causes and effects, antecedents and results —but starting from initial conditions that he himself has previously defined, and that therefore "rule" the results. In fact, stating the initial conditions has the natureof stating the hypothesis of a "law," valid for a (possibly large) class of events.

The same cannot be said to any extent of analysis of biological systems, notwithstanding the fact that the biologist strives to construct his own working hypotheses; the reason is that the initial conditions can never be stated with sufficiently reliability. Therefore, the biologist is left with the analysis of specific, scarcely controllable events, and in most cases he considers himself lucky to reach a statistical appraisal of the observables, trading knowledge of the principles for a statistically defined order, too often deceitful. Thus, physics, sharing methods and logic with mathematics and empiricism with biology, may arrive at "universal," though circumscribed truths and principles. Certainly, the domain of physical truths would be wider if analysis could start from less down-to-particular situations. The intrinsic potency in the physicist's standpoint should not escape the conscious biologist —in particular the neuroscientist— not because it could encourage a search for unification of all loose concepts within the field but for more rationally exploiting the logicomathematical and empirical paths of thought through the more thoughtful approach to the problems as typified by the physicist.

The philosophy that is inherent (even if inadvertent) in all research clearly indicates that the analytical method can be used only for recognizing chains of causes and effects, and in doing this it may also lead to discovery of the existence of "new" essences. However, to achieve this one must start from already known essences, and new ones are discovered only as objects existing in relationship to already known ones. On the other hand, knowledge of an essence is much more than merely becoming aware of its existence; it actually means gaining an "insight" ito it, into its deepest nature and together with its accessory attributes, qualities, and properties and its reciprocal relationships with the known ones. As we shall see, the motion toward the "essence" (its knowledge), and even more toward "*esse*", requires a mode of thought quite different from analysis (in the sense described in the present section) and is basically a theorizing motion. In other words, any scientific research that does not aim at a simple description requires philosophical reasoning, and the latter seeks not to be concerned with physical data (at least not physical data per se) or matter-of-fact objects and situations, since these do not represent essences but are at most only a clue as to their existence. Philosophical reasoning builds abstract structures from elements not directly linked to particular material objects and proceeds to analyze their relations within general reference frameworks (or theories). With the help of this procedure, but not as a direct product of it, new frameworks also comprising the observed properties of the present objects of thought may be formulated.

Philosophy has many common features with mathematics (many philosophers have been outstanding mathematicians) and common procedures with physics as well. The philosopher builds his own starting concept-ideas, and fully developed frameworks using the same criteria that the physicist adopts in defining the system to be analized in a laboratory experiment: through preliminary determination of the initial state (the conditions) of the abstract structures the starting definitions are given, which is another way of defining the rules of their subsequent logical development. Further manipulation of the abstract structure (the object of speculation) is, again, very similar to an experimental physical procedure, inasmuch as philosophical reasoning also probes the properties of its object through other objects of thought and is concerned with their interaction, using throughout the rigorous logic and criticism of scientific reasoning. The philosophically minded scientist does not need to use symbols to indicate the objects of investigation (although single words may be used for representing whole theories), a fact offering a certain advantage, since in the initial stages of reasoning there is no need for creating precisely defined symbols. However, the absence of a clear-cut, universally accepted system of symbols is actually rooted in the peculiar —though most disturbing— feature of philosophical reasoning, the fact that by their very nature the symbols concerned cannot easily be translated into sets of universally accepted definitions (as opposed to the case with mathematical language). Each of them requires elaborate descriptions and explanations because different aspects of the same concept do not appear with equal clarity, and their reciprocal relationships (and logical links) may not be plainly understandable to everyone.

It is no wonder that the different characteristics of reasoning have always tended to generate confusion among scientists. The science of knowing, epistemology, is concerned with determining the strictly logical definition of relationships between objects to reduce as much as possible the uncertainties inherent in language and to keep reasoning within rigorously demarcated tracks, as characterized in mathematics and physics. What matters here, the kernel of philosophical reasoning, is that the starting point is always a fully developed theory as general reference framework for all ensuing analytical procedures. In turn, analytical research characteristically follows continuously dichotomizing paths, since each object —each essence— used in reasoning must itself be investigated within the same or another reference system. The general philosophical procedure, dealing with objects not mathematically defined, should be followed by neuroscientists inasmuch as every empirical observation needs first to be framed within a theory. Science proceeds more quickly and effectively if it first produces theories, later to be put to test, and particularly if the tests can produce simple answers such as "yes" or "no." Much more ingenuity is required in formulating the right questions about origins than simply recognizing the sequence of relationships in the endless causal chain of physical events. In the extreme, a wide-embracing theory can always be put to test with a few key experiments —even a single one, provided it is consistently reproducible (although there is much doubt as to whether experiments may ever be a real test for theory).

It cannot be stressed enough that in every region of science thought has the dual aspects of theorizing and analyzing. What should not be questioned is that we may proceed with epistemologically correct procedures only from the formulation of a theory, then analyzing every aspect of it, testing its predictions, and thus eventually succeeding in explaining the particular events within a general framework. Generally in science, nonstatistical, empirical "backward" prediction or —what seems simpler— "backward" recognition of the logical path, is epistemologically forbidden because the direction of reasoning cannot be reversed as an uninterrupted, consequential logical procedure from particular or contingent to the general framework, i.e., from experiment to theory

3.2.1 The Problem of the Fine Structure of Thought

The inherent asymmetry in the logical paths of thought strongly suggests that the "fine structure" of reasoning is different in the two directions, theorizing and analyzing. If only at an ideal point, theory and analysis coincide, for the latter must take its origin from the former. We should first ask whether the elements upon which theoretical reasoning is based differ from those upon which analytical reasoning is based. The elements building the ideas and the ideas forming thought do not differ in nature between analyzing and theorizing procedures. Therefore, the problem is better considered at the level of the neural organization of thinking, i.e., of the finest or simplest neural procedure used by the eidetic subsystems. As an initial hypothesis, the fine structure of thought can be represented by the elementary operation of comparison, which was viewed above as the fundamental procedure of the neural function in general. The same comparative operations can be seen as fundamental for the eidetic function (as we have actually done so far), since on it are founded the measurements of congruence-coherence among elements building ideas and among ideas building thought. Of course, we must ask whether the comparative operation is sufficiently representative of the fine structure of thought. The unit of operation representing the fine structure might well be another, regardless of its degree of complexity, provided it withstands a critical test as basic building block of the eidetic procedures. It is also possible that comparison is not the sole procedure, but that other operations, as fundamental as comparison, together with the latter contribute to define the structure of thought. If so, we must discover what other principles of organization of neuronal activity are involved in the fine structure of thought.

Literature on this subject is extremely scant. Only the linguistic aspect of thought, with its psychological and philosophical features, has been subjected to extensive analysis. Language would have to be considered here if our purpose included the expression of thought through the "structured" language of words, ideas, and sentences; it cannot be denied that this aspect may offer a useful approach to understanding thought as organization, also in neurophysiological terms, but our viewpoint is different. We shall not consider language as the "expression" of thought, i.e., as originated from it or representing it —but rather the reverse: as the basic nature of the eidetic process we find an "internal" language from which only secondarily we may more or less faithfully express our thoughts in the structured language of speech. The internal language —and we shall see in what sense— can exist before thought and without it (but we shall defer this question until Part V). The problem of the brain investigating either its own logic through language, or language through logic, cannot be solved using only logicophilosophical and epistemological tools. Although of great utility, the latter's actual bearing on the problem cannot be defined until the basic question concerning the conditioning of these tools is also solved, i.e., how fine structure and development of thought depend on particular types of brain structures and their organization. It is clear that we shall have to be more rigorous in our approach than we have been so far, because the meaning of the last sentence amounts to search for *the* eidetic structures, and upon finding these only the critical examination of their properties may give us a clue about their conditioning nature, ways, and development of thought. A less than rigorously critical approach may lead to the acceptance of assumptions about eidetic processes —formerly numerous only in nonmedical literature but now unfortunately in the medical as well— that do not sufficiently take into account the properties of the physical structure of the nervous system and its organization.

Returning to our earlier question about the fine structure of thought, it is clear that nothing would prevent us in principle from taking the comparative procedure as representing the tool for constructing reasoning, i.e., the sought-for fine structure. The question whether alone it can account sufficiently for all the mechanisms of eidetic processes cannot be answered at this stage, although our approach presumes comparative operations to be at their basis. We are still unable to ascribe the function of comparison to any particular (i.e., ad hoc) structure or circuit, or synaptical arrangements at the level of neuronal membranes. The reason is that we always tend to include in the definition of this term a number of different operations, however having the common feature of being performed on variables (activities or their functions) of two structures, so that the two may be confronted in some way, and some measurement of their reciprocal relation may be performed. The actual neural operation (and the related circuit to do it) probably has nothing in common with the physical systems (and their corresponding principles of operation) that we use nowadays for performing comparative procedures and that mathematically and physically may produce quite different results. For instance, a mathematical correlation procedure, either statistical or not, almost certainly is not implemented in neural circuits in the same way as in a computer; neither is comparison probably performed in neurons as in physical contexts. Conversely, we know for sure that the neurons have ways of working —fundamentally exploiting the demonstrated, non-linear, logical properties of synaptical arrangements on their membranes— to perform operations giving results "practically" with the approximate function of a correlation or a comparison.

Therefore, in our search for the fine structure of thought we must recall the discussion about feedback. As we may understand it, feedback is at the same time a principle of structure expressed in a type of circuit and, strictly linked to the characteristics of it, of function, i.e., it represents the main "unifying" principle. By accepting the hypothesis that in a feedback circuit the elements interact with each other to form an idea at the very moment that feedback is closed, we accept the existence of a fundamental eidetic structure implemented in neural circuits (of greater or lesser complexity but always with the same essential and simple significance). This may also be considered as a fine structure of thought, and we should examine how it is related to that other proposed fine structure, comparison, which in turn is the principle of structure (or of different structures) and of function. There are perhaps several different possibilities for the joint action of comparison and feedback. We must consider that comparison is a principle of measurement —and therefore also of judgement— inasmuch as it compares quantities (intensities or extensions) or qualities (for their similarity or congruence-coherence). Both the quantitative and qualitative comparisons are necessary to the eidetic function. On the one hand, the nature of the idea may result from features of intensities or extensions of the constituting elements. On the other hand, every qualitative judgement has its neural basis in the appraisal of "integrated patterns" of measurements (see *PTN*), thus not resulting from comparison alone, although the individual elements of the pattern may be represented by the comparative measurements. Accordingly, feedback is a finer structure as far as thought is concerned, since it embodies the unitary abstract structure, the "idea," from which thought is developed by assembling ideas on the basis of their congruence-coherence. Comparison, on the other hand, is a finer structure as far as the idea is concerned, since it assembles the elements by comparing them to each other. However, we might put this better by saying that comparison operates on the idea from the "inside," by working on its elements in the process of idea formation, while feedback operates on the ideas from the "outside" (or, on the thought from the inside) by assembling them in the process of thought development. Therefore, the true fine structure of thought still consists in comparison, the principle which sorts out the elements of an idea and the ideas as elements of a reasoning.

3.2.2 Comparison as Fine Structure of Thought

Following the arguments of the preceding section, comparison is the finest structure of thought that we can identify so far. Its role in the eidetic function is instrumental at the stage of formation or definition of the idea by comparing the elements entering it and measuring the degree of their congruence. Comparison can be viewed as a kind of mechanism sorting the elements to be introduced into the idea and, after the idea has been formed by assembling the ideas in the process of thought development. In the assembling procedure, comparison is instrumental because every step requires a decision-choice for one idea over against others on the basis of best congruence-coherence. It might be pointed out that the measurement of congruence-coherence necessary for judgement is actually something neurologically different from the comparison itself, which is the step logically preceding measurement. However, the two procedures seem so indissolubly bound that there is no point, at this stage at least, to consider them as separate merely for the sake of epistemological precision. Comparison is instrumental in another aspect of the reasoning process, i.e., in the recognition of the cause-and-effect chain, the typical procedure of logical-scientific thought. As we shall see below (Chapt. 5), cause and effect are necessarily linked by congruence-coherence. Cause-and-effect links are recognized through the judgement of logical interdependence inherent in the laws which hold matter together through the interactions among its elements, and represent essentially a description of the dependence. For this reason, the analytical procedure is that which most takes advantage of the assessment of cause-effect recognition, as the method best fit for discovering the hidden and finer mechanisms of physical inanimate and of biological events.

Quite the contrary is the case of the eidetic theorizing process. At first sight, comparison does not seem to play a comparably instrumental role here as in analysis, since theorizing thought, by definition, does not follow the cause-effect chain in the direction of ever finer particulars. However, this is not an inherent fault of comparison as such. Comparison is perfectly suited for performing judgement on congruence-coherence in either direction. Without anticipating the arguments to be developed in Chapt. 5, comparison and judgement do not prescribe the thought direction (the mode); they only consider two propositions as to their logical fitness. Apart from the very particular case of pure mathematical reasoning, there is not a unique, logical antecedent for every (probably also not always in the case of pure mathematical concept-ideas), unless the initial conditions are rigidly established, and the system is thoroughly isolated, as could ideally be reached in physics laboratory experiments. In every type of open system several possible causes may interact, and it is generally impossible to point to only one of them as the unique cause of a particular event-effect. The situation becomes much more complicated when the initial conditions are not known to any sufficient extent, such as is generally the case of biological systems, since in them the number of variables and degrees of freedom may be exceedingly high. In the latter situations only statistical correlations (when statistics can be applied) between effects and supposed causes may represent a second-choice substitute for more exact definitions of the real, necessary cause-and-effect relationships.

At this point, it must be clear that the impossibility of logically retracing the path from effect to cause through a comparative procedure is not a fault of the principle of comparison but one only of the use to which we want to put logic. In fact, for a Maxwell's demon endowed with knowledge of the initial conditions of all variables, as well as of all the possible causes and modes of their reciprocal interaction, finding the right path from any particular event to its actual cause, through the maze of dichotomizing paths, would only be a matter of representing the whole situation as a system of purely mathematical, necessary relations, from which would flow the necessary and inescapable solution. In other words, in order to be able to reach the level of the causes when starting from the level of their effects, knowledge extending well beyond the single event is a prerequisite. Without this previous knowledge, logic alone is powerless; the logical hiatus is determined by lack of the necessary information. However, many modern biologists claim that a fundamental uncertainty, as that existing at quantum levels, is also inherent in biology, and that this always prevents us from logically reaching certainty of cause-effect relationships at some fine level of the structure organization.

In the logical process of recognizing cause-and-effect links we may find the key for unifying concepts that are still considered separately, such as comparison and operator, and for ordering them into a more comprehensive framework of relationships together with the concepts of transformation and nowledge. We must not forget that a judgement on congruence-coherence (stemming from comparison and representing mind's interpretation of the result) implies the recognition that one thing has some type and some degree of similarity with another. However, since one cannot say whether two objects actually *are* and not simply *look* similar in the absence of information about their inner structure, i.e., without understanding them in their essence, this type of essential comparison implies preknowledge. If the two similar objects are not mere examples of the same thing, the differences between the two, which are the object of quantitative assessment, actually represent the amount of "imperfection," or of nonrealization, in the process of an ideal transformation that would make one object exactly identical to another. Although we have not yet given a clear definition of the concept of transformation, we can say that transformations follow rules defined by their operators (a concept already encountered), and in this way the structure of the comparative mechanism for detecting similarity can hardly be distinguished from that of the operator which would ideally operate the transformation of one object into the other. If we can think of the objects as being physically stored, coded in some stable but retrievable form in whatever type of memory, and could compare them, the mechanism of comparison would still be necessarily tied to an operator of object transformation, which in this case would be matching templates of patterns. (However, since there is no acceptable demonstration whatever, we cannot accept that such a physically coded, stable, and retrievable memory can exist for previous experiences, neural events, or objects such as abstract structures representing concept-ideas.)

There are further reasons why we must maintain the view that comparison should not be thought of as being different from the operator of transformation. Even if such a stored memory for neural objects could exist, it still would be extremely awkward to picture, by some stretch of imagination, how material structures, such as the physically memorized ideas, could be compared with one another. Neural operations must be used for the comparison, that is, essentially nerve impulses, which implies that the stored information must be retrieved from its materially coded state and given its former "fleeting" form, that of nerve impulses circulating in eidetic circuits (that which it had before being stored). If the procedure were arranged otherwise, since the operator (and therefore the comparator) has the fundamental nature of a neural variable, however complex, it would be like using wind for comparing stones. Regardless of whether we are reading the information with our eyes or with an internal operator, materially coded information must always be transformed into neural symbols, the nerve impulses being the only symbols that can be used for further elaboration. Therefore, it seems that if we are to compare the result of a transformation with a previously stored model, the model itself must regain the same nature of the transformed object, i.e., the abstract structure (though physically implemented in eidetic circuits) that it had before memorization. However, then the procedure of comparison is necessarily performed on abstract structures —peculiarly fleeting objects— and there is no way of holding these objects firmly for the time needed for comparison or for as long as we want to examine at leisure their differences. Therefore, the comparison is a process that must be renewed continuously until a decision is reached. Accordingly, the actual mechanism of comparison can only consist of the continuous transformation of one abstract object into another, i.e., it can be identified with the recognition of the agent of transformation, the form of the operator itself, which alone describes the differences between the objects (we shall return to this below).

3.3 Theorizing Thought: Cause-and-Effect Relationships

Leaving aside, for the time being, any further consideration of comparison and transformation, let us return to theorizing thought. When mind theorizes, reasoning is aimed directly at developing a general reference framework to explain and predict all observable, contingent events. We maintain that reasoning cannot logically reverse its path and retrace the origin of things through an unequivocal, uninterrupted recognition of the chain of effects and causes —that origin which is part of the reference framework, i.e., of the coordinated ensemble of principles and laws governing the whole system. Therefore, we must ask how a theory comes to be drafted, that is, through what mental methods do we reach our goal. We have broached this question repeatedly and have always left it unanswered.

Ultimately, in considering the question of theorizing mode, we should be able to find the unique primary cause of all the possible events in the world. Even if we truncate this search well before that primary, all-originating cause and content ourselves with a partial, although sufficiently comprehensive framework for only one large class of phenomena, such as those subsumed under brain function, our aim still seems much too ambitious. Restricting the range of the search for originating cause helps us avoid discussions about the unique *primary* cause but still does not avert the danger inherent in teleological reasoning. A theory, the more general it is, should never shun teleological arguments. The intention to reach a goal is put into act through a series of means (the actual objects of our investigation) that carry some hopefully recognizable mark of that intention, and therefore of the goal, since the means need to be suitable for reaching it, hence strictly coherent withit and reflecting the intention itself. Note that the reasoning is not changed when in place of a teleological development one assumes a series of mutational changes, even more or less casually determined. In fact, the present state of the brain structure function has required such changes, and the whole procedure still represents the "necessary" steps through which the brain development "had" to go. Thus, we still are in the search for cause-and-effect relationships, regardless of how we label them.

Here we need to introduce the idea of preknowledge. Let us return to the Maxwell's demon in the form of a purposeful operator sitting at the origin of the causes (and assuming no chaos-ridden mechanism); by necessity he would be endowed with preknowledge of the effects of its operation, i.e., of the type of order eventually to be reached. This I believe the crucial argument for recognizing the set of principles, the framework for the largest possible ensemble of events or of ideas. Perhaps we too would require preknowledge to detect the overall framework comprising, if not the primary, the original cause, at least a large number of causes for as many classes of events. Put in this way, the question is not only rhetorical but unanswerable, for it is altogether preposterous (however, we shall see below that it really is not). We have said that analysis, with its comparison procedures, always works in the direction from theory to facts, from general to particulars. That is, having drafted an initial, rough overall picture —something more than a mere hypothesis— we proceed by analytically verifying logical consequences, one after the other. At each step, we generate a judgement on congruence-coherence between the elements of the framework (the general principles) and the particular items which should be explained by the former. The things seem to proceed logically only from theory to particular, notwithstanding the fact that we may represent the cause-effect as a symmetrical relationship, as with Newton's law of motion (which is ideally symmetrical); however, as the case of Newton's law, we are confined practically to unidirectionality. We shall see a possible escape from this constraint.

3.3.1 Deductive and Inductive Reasoning

Although we wish to maintain that the comparative procedure is as fundamental in theorizing as it is in analyzing, it is less easy to envisage in the former than in the latter. However, let us attempt to do so. The thought process leading to theory is based on the necessity to account logically for a number of general characteristics of the system, that is, a logical way in which, given A, B, C,...N as the system's properties, we may be led to theory X. Theory X must comprise a much greater number of properties than those in the original observations or assumptions (A, B,

C,...N). This means that a theory cannot depend merely upon specific data; it must also apply in the case of data yet to be discovered and must enable one to predict the outcome of possible experiments and observations other than those that gave rise to it.

Generally speaking, analyzing and theorizing procedures are associated, respectively, with deductive and inductive thought. While the meaning of deduction is rather clear —i.e., a logical derivation from antecedent to result— this is not the case for induction, which is not reversed deduction but an altogether different procedure as we shall see below. Induction, according to its generally accepted meaning, is a method of reasoning leading from particular data to general laws. We shall take a different view, however. The argument using the Maxwell's demon possessing preknowledge, or prescience, was not out of place. What one does in constructing a theory is to assume, first, that all the observable data are the outcome of some process according to the principles or rules of a general reference framework, in other words, that a theory actually exists which unifies them. The second step consists in using the known facts (data), as the Maxwell's demon uses preknowledge of goals, but in our case with the special connotation of what is to be reached. To be more precise, we see the data as the logical consequence of some elemental ensemble of principles and rules governing the system, and we want to reach them by finding the logical link from the unknown to the known.

In the case of the somatic nervous system, simple previous knowledge of structure and functions has helped to outline, for instance, "theories" of how the muscles are moved and organized in purposeful actions. This ensemble of knowledge and purposes, however, does not produce "general" theories but little more than tautological reasoning, the reformulation of the already visible and tangible. With this, we are still producing not theories but working hypotheses —merely introducing some order among the observables by the mechanisms binding them. No one denies the usefulness of this procedure, since a vast amount of today's research is receiving increased order through these minute steps. On the other hand, if we want a true theory, one that accounts for many more observables than have been provided by actual knowledge, i.e., for a generality of classes of events, we must not merely collect observations but manipulate these in our mind. We must make them objects of thought, abstracting from them their essence, what they actually represent within a larger ensemble of observables, and use this abstracted matter as preknowledge and purpose.

Thus, abstraction is understood here as the only process that can lead us to the root of the meaning of an object, to its essential nature. Through this we also arrive at an understanding of the essence of the "true" theory, i.e., the logically ordered ensemble of the ideas representing the relationships among the essences. Therefore, theories cannot consist simply of the principles which operate "mechanisms" (such as the working principle of an engine); the latter still represents little more than the mechanics of operation and thus a class of mental objects substantially different from those forming a true theory. If the abstraction procedure is pursue far enough, i.e., if progressively higher levels of abstraction are attained, the initial data are lost from sight. Mental or experimental testing of the theory can reproduce the logic of working hypotheses and the level of the actual data. The process of abstracting-theorizing can be tested at every step through this analytical procedure, with its own measurements of congruence-coherence; the fundamental point is that the theory must be tested against data different from those that gave rise to it, even those of other classes.

From these observations one would draw the conclusion that so-called induction (inductive reasoning, in the common sense) is largely an empty word. Applying the inductive procedure is justified only if one is seeking restricted hypotheses starting from restricted knowledge. The procedure must therefore be supplemented by another. In utilizing experimental data to advance explanations about an underlying phenomenon, the investigator does not need to assume the outcome of the experiment to be the direct, immediate representative of an underlying specific law. Rather, disregarding the paucity of specific knowledge about the object, he tries to "jump" much farther backwards, to the possible point of convergence of lines of thought from other regions. In fact, the abstraction procedure —the kernel of theorizing thought— operated on the objects and their relationships can only terminate at some general statement about an ensemble of principles, so general as to constitute the point of arrival also of other lines of abstractions in other regions of science or of mental experiences.

At first sight, it would not seem to be a different approach from the one previously discussed and rejected, but actually it is different. The crucial point lies, as before, in the impossibility of a real backward vision, i.e., of retracing the path from effect to causes. However, for the unknown principles regulating the transition from antecedent to result —representing the unknown to be unveiled— one now substitutes the principles known to be at work in analogous situations. The key element here is clearly the meaning of "analogous." This type of working hypothesis has been revealed to be very fruitful, since by taking the place of a real theory it provides at least a partial, substitute theory or some other proposition of more general value than the mere reformulation of the experimental results. The analogue-working hypothesis assumes a mechanism as probable on the basis of previous experience with more or less similar cases and proceeds by submitting the hypothesis to test in the usual analytical way.

The theory developed here (and in *PTN*) abstracts from existing knowledge about the physical world and the brain and seeks to account for phenomena in the somatic and autonomic regions of the brain as well as (hopefully) in eidetic processes, in the same terms as used for physical phenomena. In other words, the principles which account for the motion of physical bodies, when submitted to such abstraction, should also account for the motion of fleeting, abstract structures such as those of mind.

Thus, the very choice (on probabilistic bases) of an analogical explanation is, to repeat, necessarily based on previous knowledge, even if incomplete and limited. In biology there are innumerable examples of such situations because many vital processes, although pertaining to different structures, are caused in essentially the same ways, as if nature operated with only a limited number of fundamental procedures (this actually is the hidden assumption, so general that it may itself be developed into an independent theory). Thus, nature would still achieve a great variety of results by operating in a restricted gamut of ways upon a great variety of different elements. However, even if the basis of this reasoning is a probabilistic consideration, the reasoning itself is not guided by statistical evaluations, rather by the view that analogies are widespread, i.e., similarly organized ensembles of elements. The procedure is extraneous to a mere statistical appraisal of data because it stems from a fairly well-defined framework comprising fundamental principles which regulate physicochemical phenomena of inanimate and animate nature. As was said above, for this type of inductive reasoning there must be a full body of available knowledge to be exploited for the purpose.

Thus, the more elemental, yet realistic way of attaining a backward knowledge of the principles of transition from antecedent to result does not consist in trying to directly inspect them backwards —an impossible feat— but in making recourse to existing information on other instances as guideline. It is obvious that mind is continuously rehearsing knowledge (that is, information previously collected) and uses it; however, in the case of the analogical procedure, the information refers to matter not specifically related to the object of concern but to another, more or less closely related one. This means that the investigator should possess a much wider background knowledge than strictly that pertaining to his specific field, and that he must be able to recognize the relationships to even far distant fields.

3.3.2 Theorizing by Analogical Reasoning

Nowadays, because of the vast amount of data gathered in all fields of science, analogical reasoning represents by far the most preferred shortcut procedure; this is developing more rapidly than direct, progressive abstraction. Analogy is used for exemplification purposes and is based on congruence-coherence comparison between whole situations; it is peculiarly suited to both theorizing and analyzing modes. In both instances, analogical reasoning provides comparison to something already wellknown, either as an explanation of actual data in analytical processes or as a more or less comprehensive hypothesis in theorizing. However, few investigators are aware of the fact that in the theorizing mode analogy is useful only if the data or the contingent situation is abstracted precisely at the level suitable for comparison with the other situation, different but already known in its principles; only then is one justified in using the latter as an explanation of the former. This means that analogical reasoning can begin only at a certain stage in the process of abstraction, when the abstract significance of the situation suddenly suggests to the mind echoes of another one, consisting of an organized set of events, ideas, rules, or principles. The comparative procedure, with the corresponding measurements of congruence-coherence, is the operative bridge linking the two fields by establishing what they have in common. Through comparison, analogy may eventually reveal the two situations as two different aspects of a single set of laws. Thus, analogical reasoning would seem to be a common substantial component of theorizing thought. However, it may exhibit different degrees: from the crude "strike in the dark" of a mind quickly responding to faint echoes of familiar characteristics to the rationalized recognition of incontrovertible analogical similarity or the identity of intrinsic content between reference frameworks, when the process of abstraction is progressively conducted up to the point at which the two frameworks may be brought to coincide.

Analogical reasoning played a large part in developing the theory of neural function in *PTN*. There, the analysis of the properties of the brain afforded theoretical representation in different physical frameworks, each, however, accounting only for some aspects of the cerebral function and not for others. The consideration of a relativistic framework, on the other hand, had at the start its sole justification in the experimentally observed difference of time interval judgements in different states of the brain, "echoing" a relativistic situation. Although explanations of this particular phenomenon might also be reached on a different basis, the relativistic framework was able to account for other phenomena as well, such as the time covariation in the brain space (the brain *Eigenzeit*) and the required invariance of judgements and of the brain functions in general; it provided, furthermore, a common basis for explaining the interindividual differences of behavior in the same situations. Other theories, at least so far, are unable to provide a comparably wide basis of justification.

Even if analogical reasoning may sometimes be instrumental in theorizing thought, it is by no means a necessity. The process of progressive abstraction alone can delve ever more deeply into the essence of objects (things, events, ideas) and must be considered *the* basis of theorizing. Through it one may reach a general set of principles even when no other set of preknowledge is available for comparison (or for accelerating the procedure). This seems to have been the case of the main scientific theories from antiquity to the middle ages (from Aristotle to Thomas Aquinas), even up to those in our times representing the foundations of physics —relativity, quantum mechanics, and beyond. Surely, Einstein was able to take advantage of knowledge of non-Euclidean geometry, the theory of continua, and tensor mathematics (even if the latter had to be further developed), but the theory of general relativity was not derived from these nor was it a product of analogical reasoning. Other theories, such as quantum mechanics, took advantage of Planck's ideas and of the known principles of statistics (also further developed, and giving rise to new, more suitable principles) and developed the theory of operators and the mathematics of matrices. Quantum physics is not the logical derivation of previous knowledge, in the sense of necessary transition from antecedent to result. In the development of their theories the scientists were not guided by the iron law of consequentiality inherent in the search of the mechanisms. They abandoned previous, well-established tenets, stripped current physical doctrines of many uncritically accepted assumptions which could not find experimental confirmation, and proceeded to the essence of the phenomena by the mere force of the abstraction procedures of mind. Preknowledge was used not as starting point but, as we have said, as already available confirmation that the basis of the new theories were not illogical or unsound constructions but actually lent support to the prediction of larger classes of facts, which also included the already known ones.

Thus, the path of reasoning was independent of preknowledge. But, should we conclude that the role of preknowledge is secondary in theorizing? Certainly this is preposterous. Without preceding knowledge, Einstein could not "immediately" recognize the truth of his own conclusions —long before a test could be made of the curvature of space-time, which could not be accepted beforehand without a preceding description of non-Euclidean space. In the development of a theory, preknowledge *may* be the source of echoes but actually *is* the system of pillars assuring those points of reference that may combime the new with the previously existing theories and provide the elements to support further development.

3.4 Abstraction as Basis of Theorizing Thought

Theorizing thought, although starting from the same type of elements as any other thought, uses them in a different way; we might say, it uses a different heuristic procedure of reasoning. In the very general case, whenever confronted with a goal, the terms and the characteristic properties of the elements are identified and placed in relationship to each other. In doing so, the elements of reasoning are given the same importance since each contributes to defining the goal and the way to reach it, and each element is assumed to be as necessary as every other to the completion of the task. Theorizing thought seems to work in a different way. Again, comparison of objects is a fundamental procedure, but not the ultimate, principal one. We have said that the principal operation must be considered the "pure" abstraction process, that which alone characterizes theoretical reasoning. Using a figurative expression, one might say that abstraction moves in a plane "perpendicular" to other reasoning procedures, since it typically goes further into the essence of one single object of thought and is therefore a procedure that does not compare different objects. Abstracting the significance means to be concerned with one object and attempt to gaining the most profound and complete knowledge of that object alone.

We need to be more clear about the role that we assign to the abstraction procedure. The "most profound knowledge" of one single object might be seen in different ways. Let us say that the abstraction leads us first to the definition of essence, i.e., the set of properties that make the object what it is. However, this definition of essence is still rather ambiguous. If we progressively modify the complex of qualities (by adding some, for instance), we still have an essence, albeit a different one; we still have the kernel of the previous essence but now with some property more. Let us introduce a crude example. Among living beings we can isolate the species man from the species cat, each having an essence based just on that set of properties or qualities necessary for the outward distinction. We have, then, two essences which contain all the information (descriptive qualities) needed to make man and cat what they are. However, "what they are" cannot be understood in an absolute sense, i.e., in their more essential nature, but as "what they look like." We may now consider more closely the essence of man, adding to those external properties another property that we believe to be fundamental and better characterizing man with respect to other living beings (such as the cat), for example, the intellect. We may further add the quality of obedience to ethic principles and further still qualities referring to social life. In this adding ever more properties that we believe to befit the essence man, we shall at some point exhaust all the available properties.

This view leads us to two conclusions. First, it is clear that knowing all the properties and qualities of an essence, that is knowing an essence completely is an impossible feat, even when concerning only the material aspects, since matter continuously reveals new, unsuspected, and sometimes incomprehensible aspects. In the case of living beings, their essence remains completely obscure; besides the problem of purely material (physicochemical) properties, the essence of life itself remains unknown. It follows that an essence, in the only way in which we may know it, is reduced to nothing more than an incomplete list of qualities. Furthermore, the essence so conceived is not a fixed reference and lacks something "hard," an existent substance. Failure to recognize the fundamental difference between essence and *esse* has caused the most unfortunate misconception that has befallen natural scientists (but not only them). As noted above, scientists who investigate physical nature are easily led to the conviction that reality is bound to essence alone, because the individuality of the investigated objects is based on the incomplete definition given of their essence, and the relationships among the objects are no more than relationships among their essences, without taking heed that the latter are simply relationships among incomplete sets of properties. In this way, however, each set of incomplete properties defines something which exists only in relation to other sets, not by itself. Investigating only incomplete essences and the relationships among these always limits the investigation about what constitutes "quality," about the attributes, not about the "reality," not about the entity that alone is endowed with that reality and at the same time with those attributes, but which has many more —perhaps infinite— other attributes. Ignoring the difference is not harmful until one misinterprets the essence as the true *esse*, as when in the abstraction process any one of the several grades of the essence, i.e., what appears, what is relative, is misunderstood for the ultimate nature, the real *esse*, the nonrelative, the firm, immutable reference.

Comparison therefore enters into theoretical reasoning at a later stage, when after abstraction —or, a "reduction to essence" up to some level— the individual objects are seen as isolated sets of elements, as essences. At this stage, the various essences or more precisely what results from the reduction to essence up to that stage, can be put in relation to each other. Thus, what the comparative measurements (in this case, congruence-coherence) reveal are the observed relationships between essences; the further the process of abstraction has proceeded, the more fundamental are the relationships that are detected. These relationships are themselves abstract in nature, that is, they represent the principles by which the essences are linked. What matters to us is that the comparison at this "reduced" level brings to light relationships between reduced objects, perhaps originally also belonging to different classes or regions not reciprocally related (certainly not at first sight) at the level of physical reality or raw (not sufficiently reduced) ideas. The full set of relationships appears only when the objects are seen after reduction; alone this procedure offers the opportunity for a comprehensive view of how various objects are reciprocally related, represented in essential terms. (Nevertheless, we must not forget that the essence thus reached is only a representation, what "clothes" the reality, what the reality "appears," not the reality itself.)

Attaining an overview of various reduced objects (i.e., ideas, problems, questions) requires that the objects in various fields be understood, not in their technical details but in their intrinsic essence. This is to say that theorizing thought is based on the integration of knowledge concerning matter —physical, chemical, and biological— with other sort of knowledge of no less scientific nature, although not strictly logico-mathematical or matter bound, such as broadly philosophical knowledge. This concept is noted repeatedly in *PTN*: in no field of science can complete knowledge be attained if the individual problems and the objects of that field are not part of a larger, more comprehensive picture, comprising them all, together with problems and objects of other fields. In mind, physical and nonphysical realities are complementary, and all are necessary for a complete knowledge of each one of them individually; confining oneself in either of the regions amounts to eliminating from mind half the horizon of objects that may come to constitute it. According to a generally accepted fundamental principle, the best behavior is exhibited in all circumstances and situations by those subjects who have previously experienced the most varied living conditions. Through this experience the subjects learn a wide range of situations, which constitutes the necessary first step toward integrated knowledge and thus toward the expression of appropriate reactions. The same is true in the case of nonphysical situations, such as when mental activity is kept engrossed with manipulating abstract structures from every possible type of problems. In this case there is also teaching and learning, a self-teaching and a learning concerned with the results of the "reduction to essence," i.e., with the essences and their reciprocal relationships, as they are detected through mental experiences.

Therefore, the capability —besides the logical possibility— for building a theory, more than the outcome of specific knowledge within one field, is the original product of the elaboration of a substrate of knowledge enabling mind to see through the appearance of things and to abstract the elements not exclusively pertaining to a restricted class of problems but fundamental to many. The theorizing capacity thus depends on self-training in the theoretical mode of thinking, i.e., in abstracting the essence of mental objects, and the whole procedure is therefore creative since the essence of nonphysical objects may be seen differently by different minds, i.e., in this respect is a creation peculiar to each individual mind. The procedure is creative also inasmuch as it uncovers relationships not deducible from the direct comparison of objects but graspable by mind only after reduction to essence. For these reasons, Dubarle speaks of speculative spontaneity (see sect. 3.1), the activity óf a free mind, the mind of a man not constrained within the boundaries imposed by the law of consequentiality. In the realm of the consequential and deducible the path of reasoning may be largely predetermined, even if the scientist is not aware of it. However, the predetermination is not inherent in the nature of the objects themselves or intrinsic in their relationships —which are multiple and therefore not uniquely determined— but is a function of the level at which they are considered, the level of the bare appearances, of the visible and measurable with the physical, standard measuring instruments.

At this level one can only proceed in a strictly consequential order from antecedent to result, from cause to effect, from before to after, which means exploring extensively the detailed machinery of things, i.e., the relationships linking visible and measurable objects. By this procedure, however, one cannot explore their essence (which belongs to the abstract realm and requires motion into the depth of each single thing), and therefore the detection of essential relationships, constituting true knowledge, is barred. Through progressive abstraction one eventually arrives at the hidden, necessary, and not thought-of relationships between reduced objects, therefore also at predetermined paths of reasoning. At this point, however, pre-determination no longer constrains mind's reasonings but actually constitutes the paramount achievement possible. In fact, the relationships uncovered at this most reduced level represent the fundamental laws of nature, the laws which bind the multitude of its diversified objects, ruling at one time not only the material but, and principally, the nonmaterial world. The predetermined character of consequential reasoning, on the one hand, and of abstract theoretical reasoning, on the other, assures us about the essential truth and the reliability of the results: at the level of the "machinery" of nature, in the first instance, and at the level of the "laws" of nature, in the second.

3.4.1 Relationships Between Theory and Knowledge

That which we are appropriating in our mind, and which enters as part of our knowledge is the set of principles which constitute the theories, not the particular data (which depend on the theory in so far as they derive from it their explanation). We have discussed the problems related to building knowledge, and we have insisted that knowledge is built upon essences, within a wide horizon of other objects and essences. The horizon comprises all the elements making every known object, such as is appropriated by one's mind but somewhat differently from the same object known (or appropriated) by another mind. The difference is hardly in the essence, since the abstraction procedure is not pushed to the extreme, but in the wider horizon in which the essence is placed. These horizons, comprising a great variety of elements all logically connected, themselves constitute the source of materials for theories. A wider horizon implies a greater number of logically connnected elements; thus, the richer the source, the more comprehensive the theories are that can be founded on it.

With this, we are actually linking —if not practically unifying— knowledge and theory. As a matter of fact, the process leading to knowledge, i.e., the integrated "vision" of the object within a wider horizon together with its attendant properties and other logically related objects requires that the abstract, i.e., logical relationships between the objects be recognized. This is a feat, however, that only a trained mind can accomplish —trained in abstracting essences and in philosophically understanding their reciprocal relations— in other words, a theorizing mind. In this sense, every theoretical science involves philosophy. For the same reasons we note in *PTN* that scientific knowledge —a knowledge of essences and of their relationships— is limited by the finite capability of the human mind for correctly understanding complex constellations of relationships which are dependent on the intrinsic properties of the sets of abstracted objects and by no means restricted to those that mind can understand (we return to this important point below). Positivistic science claims absolute objectivity, that is, independence from subjective opinions, inasmuch as it ideally concerns properties of nature intrinsic in its elements and inherent in their relationships; however, as a product of mind, positivistic science is actually substantially conditioned by the individual vision of nature and by the capacities of mind. Therefore, no one's knowledge of human mind is exactly similar to another's, and no theory has exactly the same connotation and formulation in different minds.

3.4.2 The Fine Structure of Theory

Let us now consider theorizing motion from another point of view. We said that abstraction, or reduction to essence of objects and relationships, is the basis of theorizing, inasmuch as a theory encompasses "in essence" the constellation of relationships. We should ask ourselves how far the procedure can actually be pushed, or, in other words, what one encounters at the end of the progressive reduction to essence (supposing this end to be attainable). It seems quite natural to answer that at the very end we should be confronted with the essence and *the* unique, absolute *esse*, the hard basis of all entities and essences and therefore of all the patterns of relationships. In other words, there would not be *a* theory but *the* unique theory comprehending in its many aspects and in all its possible developments every other possible theory. This, would also constitute "absolute knowledge." Therefore, at the end of the theorizing journey we should attain the origin of all chains of causes and effects. This also implies that every other theory can be logically constructed by starting from a more fundamental one, simply by following the logical chain of antecedent to results of the connected theories —thus with a procedure that we may call analytical (though not in a cause-effect) and using the comparative tool to check at every step the reciprocal congruence-coherence with the preceding ones. Therefore, when building a theory from ensembles of data provided by observation, mind needs to abandon the analytical procedure which has produced the data and must substitute for it the abstraction, the reduction-to-essence procedure; the law of "logical" consequentiality reenters when, starting from the absolute or from a more fundamental theory, the aim is to uncover subordinate theories.

Seeing, then, that when abstract relationships between theories are concerned, we are not dealing with cause-effect consequentiality but only with logicality, we should enquire whether at the level of theories (i.e., of theorizing thought) it is also possible to follow the reverse direction, to reach the ultimate, absolute theory beginning from lower level ones. We have observed that from a given theory other, less general or partial theories originate and it might be thought possible that by assembling less general, partial theories we can reconstruct the basic theory through a comparison procedure, measuring congruence-coherence among the partial theories, exactly as with the elements in the procedure of forming an idea. In other words, there ought to exist a procedure of theory building in which not only abstraction in itself but also comparison may be essential; theorizing would not be substantially different in heuristic procedure from building ideas.

When we consider the procedure somewhat more closely, the conclusion becomes obvious. An idea is by nature an essence —a true, albeit sometimes elementary, abstraction— therefore in this case an abstraction is attained also by assembling elements. Transposing the same concept onto theory building, which we said is founded upon abstractions, we only need the elements to be at a sufficiently abstract level (at least at the lowest, an idea). On the other hand, the elements of most ideas are ideas themselves, thus, the idea represents the "fine grain" of the theory, and since the idea itself represents a complex of relationships among its constituting element-ideas, we might consider it as an "elementary" theory. Every thought, then, that is formed by assembling partial theories, ends in the formation of a larger, more comprehensive theory, and by assembling any number of these one should eventually reach the ultimate, all-originating, unique theory. In conclusion, the theorizing procedure represents the paradigm of thought, inasmuch as every thought is made up of connected ideas. Since what we have called analyzing thought is not a true thought in the sense just outlined but the reverse of it, one can see it as a thought-destroying procedure in its fragmentation of theory —even if only a set of connected hypotheses, crude though as they might be— toward the actual data, toward the non-idea.

Abstraction is the basis of idea formation and of theorizing; consequentiality and causality are the basis of analyzing. Comparison enters in both procedures by probing the congruence-coherence at every step. However, we must partly modify the conclusion of Sect. 3.2.1, since a mere instrument, a simple comparison cannot claim alone to represent the fine structure of thought which is made up of essences; rather, in a more precise epistemological sense, we can now identify the fine structure with something more complex, that is, congruence-coherence (surely including in the procedures also comparison), the fundamental, ideal set of relationships which, as the logical plan of a building, logically holds together the elements of thought.

We should also consider the notion of models. Model building is generally understood, first, as reduction of one event or function to a few fundamental properties that can be described mathematically. Some models afford a direct, physical, and easily describable analogue, with a development subject to easy prediction or known beforehand. The predictions are then compared with the experimentally observed developments, and the judgement of congruence-coherence is taken as judgement as to the truth of the hypothesis intrinsic in the model. As it appears, the model is a working hypothesis, the sketch of a restricted field theory (a so-called theory valid for only one event or partial function) reached only through an analogical reasoning. A model requires, as it appears, a preknowledge of purposes as well as of the properties of the event or function, on the one hand, and knowledge of other areas of science providing the models, on the other. While it is perfectly legitimate to assume models in attempting to explain an event or function in terms of its mechanics, one must remain aware that the validity of models is always restricted to that purpose and to that ensemble of properties.

Thus, when one considers a larger number of properties for the same event or function, the model must be changed, and it quickly becomes impractical as a result of its increasing complexity. For this reason, model reasoning is typical of the analytical procedure, with which the mechanistic development (the pure "mechanism") of ever narrower range events are investigated. Model reasoning cannot be used in theorizing. Although the latter also uses analogical procedures, it should be clear by now that analogies are always stated between whole and comprehensive frameworks of principles —or laws— in turn resulting from abstractions which explain whole classes of phenomena or functions and are never concerned with the mechanics of particular events, functions, or actual data.

Part III

Eidetic Dynamics

4 Physics of Eidetic Motion

The arguments developed in this chapter concern what one might call, in its widest sense, the physical "drives" of eidetic motion. As with every other aspect of the eidetic function, the motion of eidetic subsystems may be viewed from different points. There is the mechanism of excitation at the neurophysiological level, i.e., the membrane synaptic electrochemical junctions, and the hodological network of interneural connections, together representing the substrate of every activity of the concerned neurons, and in a sense also the immediate drives of their activity. Although all this is indispensable, it still does not represent *the* eidetic drive, but rather the common mechanisms that set the neurons in motion, and that underlie every brain structure in every brain region, subserving every brain function. In addition to the mechanisms, we may consider the level of physical principles that govern these neural functions, and especially the eidetic function. These physical principles do not interfere with the mechanisms of excitation; instead, they are concerned with the latter's organization in a broad sense, by taking care of the integration of eidetic processes with the other brain activities and functions. This will lead us first to consider eidetic processes as a general framework of reference not only for the activity of the eidetic subsystems but also for the whole brain. As is suggested in *PTN*, the coordinate "time" of brain function space depends on mental engagement, thus making the eidetic system in some way the arbiter of the characteristics of the whole space-time of brain, i.e., of the whole brain reference framework. Analyzing the motion through the ideas' phase space, as it might be represented by the ongoing activity of the neurons, a type of Brownian eidetic motion, will present the "field" properties of some ideas. These could account for motion through the ideas' phase space since the concept of a field includes the existence of forces and lines of force, with gradients of a still indefinte, yet real, eidetic energy, representing not the common mechanism of neuronal excitation but an actual eidetic drive. However, this cannot represent the unique drive at work, since there can be an eidetic motion also in a counter-gradient direction, therefore needing a suitable operator.

4.1 Abstraction and the Reference Transformation

We must consider the source of the notion of abstraction, which is required to start thinking, i.e., theorizing. In the analyzing procedure we only need the concept of causality-consequentiality (a concept intrinsic in the structure of nature, since the latter appears to consist of an endless sequence of causally connected events). In theorizing, on the other hand, this concept is no longer adequate. We have seen theorizing as founded upon abstraction processes; however, the concept of abstraction must already be present when one first starts to abstract. In physical nature there is nothing that suggests the concept of abstraction —which requires that mind abandon the domain of the senses and delve into the invisible and otherwise perceptible. As suggested above, the process of forming "classes" of natural objects on the basis of common features has nothing in common with a true abstraction process, although it is generally considered as its basis. Abstraction does not consist of extracting the perceptible or observable properties common to a population of objects; rather, it uncovers the basic essence of one single object. Neither can we primarily (i.e., voluntarily) construct this concept from within, since the needed heuristics of the procedure cannot be known before any paradigm of the abstraction process. We can offer two guesses: a) One stumbles across the process of abstraction by mere chance, which assumes some disorderly, i.e., nonorganized, motion of the eidetic processes, or at least some involuntary motion of the eidetic structures. b) The abstraction process represents *the* natural process in some particular state, i.e., it becomes the only natural mode of thinking when a particular situation arises within, or is created by, the brain.

According to the first hypothesis, no independent stimulus or drive would be needed to start abstracting. Once some particular constellation of conditions is casually realized in the eidetic structures, the object with which mind is concerned is subjected to abstraction. From this moment on, mind can proceed by itself, through the usual judgement on congruence-coherence (or whatever other broadly comparative procedure), to detect the features of an abstract structure that may account for the nonabstracted example contingently provided by nature. Furthermore, having detected the type of relationship between the nonabstract and the abstract objects, mind may become able to replicate the procedure and apply it to other objects. Finally, by understanding the relationships among abstract objects, mind may also arrive at an understanding of the relationships among objects of different regions, i.e., at theorizing.

The second hypothesis is actually not basically very different from the first. To initiate the abstracting procedure, the eidetic structure must first be in a definite constellation of conditions. However, this constellation would not have been reached by chance but through a change of the state of mind, i.e., a transition operated by state-transition procedures, guided by the state-transition controlling structures (see *PTN*, and Sects. 1.8.2, 2.3, this volume). In this case, we would still not need an independent stimulus or drive specifically for abstracting, but abstraction would represent the unique, logical mode or procedure of reasoning in a given state of mind; it would be for mind the natural mode to follow.

Something more is hidden in this interpretation. We said above that what does not appear (judged) as logical in a given state of mind, might exhibit its logical nature in another state, for instance, when the reference framework has shifted from the region of the objects of nature to the region of their corresponding abstractions. We could state the proposition somewhat differently, saying that by changing the state the judgement on logicality may change as well. Obviously, what is changed cannot be the intrinsic logicality of the connections between objects (events, concepts-ideas in general) which depend on the nature of these things but only the judgement about these connections, an operation requiring the assessment of congruence-coherence through suitable measurements. Thus we are probably confronting two aspects of a single hypothesis. While the criterion of logicality is based on the congruence-coherence of the ideas, the measurements of the parameters on which congruence-coherence is judged may differ in two different states, i.e., in different reference systems. If in such a case the parameters do not covary, the judgement on congruence and the resulting judgement on logicality may well differ (while, of course, they remain the same in the case of covariant parameters). Thus, the question of the presence of a stimulus to abstraction, or of the suitable state-ransition control, could be unified in the drive to seek logicality, whenever logicality cannot be found by any means in the existing reference framework (conditions or state). The absence of logicality is therefore tantamount to a difference of "potential" in the eidetic subsystem. As with an electromotive force (i.e., a difference in the electric potential) naturally driving the motion of electric charges in a suitable physical system, the "logic" potential difference in the eidetic subsystem would be able to drive the abstraction process naturally. This interpretation has as a corollary that whenever a congruence-coherence is found between two objects, the abstraction process is not pursued further, at least as far as this physical, "natural" drive is concerned. Recalling the arguments of sects. 3.4.1 and 3.4.2, we are actually saying that mind starts really thinking (i.e., abstracting) only when either simple observation or the events' decomposition in causes and effects (analysis) reveals an otherwise insurmountable need of logic in a situation.

This highlights again the role of comparison at the basis of the congruence-coherence measurements necessary for logicality judgements as the fundamental brain operation. We may therefore consider this procedure as being built into the structure, perhaps as the only one to be preprogrammed (I would prefer to say genetically determined) among the brain-mind processes, not necessarily having in its mechanisms the criteria for determining logicality potential. This would account for the characteristic of "eidetic drive" as a built-in force (peculiar to the human species) to find by every possible means satisfactory, logical relationships among events of nature, or among abstract objects. In this way, the principle of comparison is the naturally spontaneous operation of human brain, although performed at different levels of abstraction. But, as a drive, this basic eidetic process is endowed with the same importance for guarding life as other, more animal-like drives such as feeding and mating, i.e., for man a basic species-preserving mechanism. These arguments support the hypothesis that the analytical procedure —consisting of recognizing the existent, matter-of-fact congruence-coherence in causality-consequentiality chains among the objects and the events of nature— also represent some kind of natural, perhaps itself built-in or preprogrammed, heuristic principle of motion of the eidetic structures. At the same time, theorizing based on abstraction would not be activated, i.e., would not become the natural motion, unless needed for judgement on congruence-coherence, which could represent the actual stimulus or drive for the corresponding reference framework transformation.

Linking the judgement on logicality at different levels of abstraction to different states (or reference systems) of the brain, entails linking the eidetic processes to the general state-transition operating mechanisms already considered. We should add some comments on this subject. Let us take this unity as a sound hypothesis until demonstration of the contrary, but with different development in time according to different situations. A first set of brain state transitions is operated in coincidence with the circadian rhythm, which determines not only the periodic wake-sleep cycle but also the oscillations in readiness for physical exertion and the overall performance observed within a given phase. However, it has not been unequivocally demonstrated that the oscillations in eidetic performance are necessarily organized as those of physical nature, with the well-known fixed periods of maxima and minima during waking and resting hours. The hypothesis of circadian oscillations in eidetic performance can be accepted as a likely to start with, but to be substantiated it requires much more in the way of meaningful tests than have been provided so far; in addition, it should also account for individual adaptation to life habits. On the other hand, the hypothesis may no longer appear likely when we consider that the eidetic processes seem to be not, or only little, influenced by autonomic functions and by metabolism, which are among the factors affecting physical performance within the circadian rhythm. It may eventually be shown that the eidetic processes are in fact affected by the oscillations in circadian rhythm, but secondarily to the changed level in general performance, more as a sort of a psyche-dependent lowering of readiness to perform eidetic work.

4.2 Eidetic Training

The eidetic function by its nature requires long and painstaking practice, one might say "training," in using the eidetic structures for developing the necessary procedures in the analytical mode and, particularly, in the theorizing, highly abstracting mode of thought, although the corresponding drives themselves may be naturally built into the structure or preprogrammed and triggered by the suitable conditions (see preceding section). In physical exercise training, the slowly developing modifications in muscular structure are the determining factor in the increase in muscular strength, but they are far less important for physical exercises based on neuromuscular coordination, the parameter actually to be improved in these cases. We may assume a similar situation for the eidetic processes leading to abstraction. Slowly developing changes may be induced in the functional-structural organization of the eidetic subsystems by tasks imposed upon it, if not continually, at least frequently and regularly; however, as constituting the intellectual function, the eidetic system in general may not depend essentially on the increase in physical determination of brain's matter —at least, there is no clear-cut demonstration available to this effect.

As already stated, one cannot accept that the logicality intrinsic in relationships among concept-ideas may be changed by training the eidetic processes in handling them, or that they may change according to mood. There is a self-consistency in nature, an autonomous and stable, extended complex of relationships among its components, including our psyche and the eidetic processes (phenomena still part of the biological aspect of man's nature). Therefore, we should see eidetic training as merely accelerating the procedures of recognizing this complex of relationships, on the one hand, and slowly adapting our criteria for judgement about these relationships, on the other. These are the possibilities for improved capability in using the eidetic structures —a facilitation of the theorizing thought processes, until they become a stable feature of mind's operation. Thus, training in eidetic activity would bring about an increasingly improved eidetic "capacity." The difficulties in this view lie in the use of such indefinite concepts as "capability for using" and "facilitation" (as far as we know, they may even cover the same neural processes). Unfortunately, we are unable to give more substance to these terms in the context of such complex functions.

However, concerning a putative training effect, no matter how evoked, the suitable stimulus for driving theorizing thought may conceivably be represented by the urge to make comprehensive hypotheses or, more generically, to solve problems in the realm of eidetics —another way of indicating the effect produced by a logic potential difference (or gradient) in the eidetic subsystem. This may be a strong factor for arousing mind's attention, not to an investigation of the relationships among objects of the senses but to an inwardly directed, in-depth exploration of one single object, an act in some way facilitated by repeated presentation of similar situations. That the brain's directed attention may be aroused by particular stimuli of high informational content (therefore, strongly attention demanding), no matter how low their intensity, is a well-known and documented phenomenon, obviously requiring facilitation (or *Bahnung*) along the path of the nerve impulse trains to account for their strong arousing effect. Likewise, it does not seem difficult to accept that also in the case of some particular eidetic tasks the attention is aroused and facilitation achieved with simple repetition. (The well-known alpha-arrest reaction of the EEG upon mental engagement is the neurological, elementary-objective aspect of inward-directed attention, and there is ample evidence that the effect may be subject to conditioning.) By assuming a change of state in some brain structures (or the whole brain) triggered by eidetic attention, we seem to fare again on firmer ground, i.e., we are assuming processes with well-known neurological bases (although not yet sufficiently clear in their mechanisms), having properties similar to the mechanisms controlling the level of alertness.

However, on second thought, this assumption may not appear necessary for a number of reasons. a) There is no demonstration that the general attention processes exert a specific influence upon the eidetic functions, besides the trivial fact of triggering consciousness on and off, allowing the eidetic processes to develop only in consciousness. b) The often-repeated change of state, by whatever mechanism produced, by itself might be sufficient to cause the irreversible appearance of some properties (i.e., giving rise to some permanent situations) such as a stable disposition of mind and the enhanced capability for abstraction accompanying it, retained for a long time or even throughout life. There is still no real answer to these questions. We shall see below (Chap. 9) that consciousness —the necessary prerequisite for eidetic activity— has other facets besides indicating a state of being and is more deeply intertwined with the eidetic processes than may appear at a first glance.

Some of the aspects of the problems connected with learning, knowledge, mind, and consciousness have already been discussed, eventually leading to the conclusion that a simple, nonspecifically increased performance capacity of our neuronal circuits could be necessary and sufficient to bring about some sustained changes of reference framework in our brain. The nonspecifically improved performance is an aspect of learning (training itself is a powerful learning factor) and of its stable influence on all other brain operations. The simple, repeated exposure to an experimental set —a training in a "situation"— is sufficient also in animals to induce the appearance of the phenomenon called "learning an experimental set," i.e. the very simple, basic generalization from a specific learned situation to other, roughly similar ones, the recognition that they belong to one and the same "category," in other words, though not yet itself an abstraction, at least a first step toward abstraction. I do not dare maintain that the arguments advanced so far meet the objections to the hypothesis that the changes of state of the eidetic structures and the change in level of alertness are controlled by the same structures. The major hindrance resides in the still almost complete lack of information about the nature and the organization of what we have called eidetic structures and their connections with noneidetic mechanisms. We cannot hope to give an exhaustive answer to these questions, but below we propose hypotheses about their general ways of operation.

4.3 Some Relativistic Hints

From the observations in the previous sections we conclude that as a result either of some specific drive or of suitable training the abstraction operations become a stable feature of the eidetic processes and a natural path along which theorizing thought develops (as seen above in the case of analytical reasoning). At the basis of this new disposition of mind is a change of state of the eidetic subsystems; however, we have let undecided whether there exists a specific state transition mechanism for the eidetic subsystems different from those operating the general state transition and responsible for the control of the general level of alertness. At any rate, whatever state transition mechanism is assumed, we must account for the possibility (also considered in *PTN*) that not necessarily all the concerned subsystems are involved equally in the state transitions (for instance, in situations of insufficient or differential training). In the latter instances some components of the eidetic subsystems would not be "at rest" with the abstraction "motion," and this fact would in turn certainly be sufficient to generate a difference in congruence-coherence between concept-ideas contributed by different parts of the subsystems (some being correctly abstracted, some not) and thus to a judgement on non-logicality within the eidetic operation. This is an example of a relativistic situation in the brain that enables us to look into the brain and to detect the relative motions of the subsystems, not through direct measurements of their proper times but, perhaps more simply, through the lack of congruence among the results of abstraction operations. Whether the result of training, or of more complex operations of the general state transition mechanisms, the whole eidetic system (or even the whole brain) is involved in the transformation; only then, when all the concerned subsystems are at relative rest with each other, can the eidetic system correctly make the judgement about logicality.

Continuing this line of reasoning, we can see that in the case of persisting nonuniform relative motions of the subsystems the process of comparison itself for judging congruence-coherence may no longer be recognized as logical and therefore not continued ay further. Again we stress that uniformity of motion of all the eidetic subsystems is the necessary basis for making a judgement on logicality of the results of an abstraction operation. To put it more clearly, in the first case the mechanism of comparison would be unable to overcome a defect in logic —that preventing the successful development of theorizing thought when the reasoning is founded upon insufficiently abstract concepts. The comparative operation of the congruence-coherence procedure can immediately detect the difference between levels of abstraction and automatically produce the signal for a judgement of non-logicality for every combination of concept-ideas. As a consequence, the comparison is cut short, and the abstraction process held up until it eventually succeeds in equalizing the abstraction levels. In this case, with the subsystems' motions at relative rest with each other, the reasoning finds the concerned concept-ideas all at the same level of abstractness, and the measurements of congruence-coherence can be performed on the properly "prepared" substrate allowing meaningful measurements and thus judgements on logicality or non-logicality, according only to the content of the ideas.

4.3.1 Thermodynamics of Eidetic Processes

The procedure of abstraction has been seen here as fundamental in theorizing thought, as that which provides the bulding blocks for constructing a theory. These building blocks result from transformation of "raw" elements —the actual data from sense organs or ideas resulting from transformation of rawer elements. Therefore, each such block is an abstract structure not directly representing elements provided by that from outside the brain (e.g., the elements issued by sense organs, measurements of some physical parameter) but the outcome of a judgement about their content, a first step in penetrating their essence. As a judgement about content (essence), it has lost direct reference to those contingent elements that may have given rise to it, i.e., it no longer represents anything material, only "relationships" among elements. If the elements are provided by material data, their relationships are only one step away from materiality, say, the idea of horse, container, etc. On the other hand, if the elements are themselves more abstract structures, say, ideas, their relationships may be considerably removed from materiality, perhaps even without any reference to it.

Thus, the results of the abstraction process are represented in different degrees, depending on the degree of abstraction involved in the elements used. This puts certain constraints upon the thought procedure, which may not be apparent at first sight. In the general development of reasoning, elements having different degrees of abstraction cannot be used; for instance, I cannot say that *my* cat is less intelligent than humans. This statement is obviously true, but it is put in an illogical way. A logical statement would describe beings belonging to the species cat as less intelligent than those belonging to the species man. Likewise, one cannot speak of ethics in reference simply to the *animal* man. Only when man is referred to as the specifically human creature endowed with the properties of intellect and spirit through which he interacts with other men, may we put ethics and man together.

These considerations return us to comments advanced above concerning the judgement on nonlogicality being founded upon a (consciously?) recognized difference between the subsystems' motions. Relativistic phenomena appear when the comparison leading to selection of the elements of thought is made between ideas at different levels of abstraction (in other words,the element-ideas used are not "at rest" with each other). We therefore have at our disposal an automatic (but also consciously monitorable) way of detecting by ourselves the state of motion within our brain through an internal observer. Necessarily, this must not be at rest with one of the motions concerned (the idea as a state, in physics a state of motion; see *PTN*, and Part IV, this volume); otherwise, it could never recognize the relative degree of acceleration, i.e., the degree of abstraction. However, if that observer could effectively judge about *all* the present possible degrees of abstraction, it would mean that in our brain we harbor a fixed reference for all motions, the observer being necessarily at rest with none of them. If we extend this argument to the theorizing and analyzing procedures —which are distinguished by a logical unidirectionality of the eidetic motion— we arrive at the same conclusion: the internal brain observer can see both theorizing and analyzing motions and can judge about them separately, i.e., is not at rest with either of them, therefore constituting a fixed reference! The last two points deserve a much more profound examination than can be afforded here.

The general situation discussed in the preceding paragraphs has an analogy in thermodynamics, the principles of which can be used for the interpretation of phenomena well beyond the physicochemical domain. The general concept of forces, the resulting effects, and the conditions of operation with which thermodynamics is concerned also apply to elements other than material particles and related fluxes (of matter, charges, energy, heath, etc.). Our case makes use of an "abstract" thermodynamics, but we are justified in so extending its principles if we retain in the more abstract context the same fundamental relationships among abstract parameters corresponding to forces, fluxes, and conditions as prescribed for the parameters in physicochemical context. Thermodynamics prescribes that at equilibrium only forces and fluxes of the same tensorial order can be considered. For us this means that the elements must be at the same level of abstract significance (the levels of abstraction associated with the different tensorial orders being different) for the conclusions drawn from their relationships to be meaningful at all. Likewise, in our reasoning about life —comprising both material and immaterial phenomena— the thermodynamic constraint must be met, not because of its physicochemical significance but for its logical and metaphysical essence.

A word of caution is necessary here. Pushing the preceding arguments to the extreme, we might feel tempted to conclude that through the refinement of the theoretical "sight" one should always reach a logical explanation of a fact or partial theory. The necessary and sufficient condition would lie simply in attaining a sufficiently high level of abstraction for each of the elements of thought, so that their ultimate (primary) common origin in a unifying reference framework and thus their reciprocal relationships can be detected. No one would likely deny a common origin for everything, regardless of whether this is through an external primary cause such as "creation" or an internal cause of development such as that of a self-sufficient nature, i.e., the mere quantitative relationships among the constituents of matter in its state before the big bang. However, explanation based on a primary common origin is of little use outside a philosophical context. A common origin of all motions and developments can be accepted as a principle, but is of no help when a judgement on likeness of two abstract objects is required. The likeness, or kinship, of different ideas must be recognized at levels of abstraction much lower than that of common origins. In other words, different concept-ideas must meet at a common origin which is not the primary common origin of everything but only the level at which we find identity on soundly based analogical reasoning (general problems of analogy are treated below.)

A further aspect needs investigation here. All the preceding arguments concern conscious judgements on logicality, and these can result only from an examination of factors for and factors against. However, one can presume that the development leading to some of the judgements does not occur in the state of consciousness, at least initially, but reaches consciousness only at a final stage, during the very act of judging. It is obvious that also in this case we are confronted throughout with rational processes, albeit proceeding partly inadvertently, possibly subserving some fundamental function that cannot or must not be tampered with by the voluntary development of thought. We can safely assume that these processes concern primarily the psychic domain, in which unconscious and conscious elements are variously intermingled and alternately take the upper hand, and in which the major chracteristically human qualities are rooted. For example, a judgement on nonlogicality may cause (or explain) the refusal of splitting the psychic unity of the subject. We can reverse this sentence, saying that the conscious judgement on nonlogicality (albeit reached partly unconsciously) may be caused by (or explained by, or be based upon) the refusal to introduce separations into the unity of psychic life. This particular judgement, in its proper essence, seems to represent a judgement on legitimacy of acts since it bears directly on psychically guided behavior. In the more general case, the judgement on legitimacy also concerns a fundamental quality of our brain, its unity of action, seen as the principle in every operation of the brain and having an undeniable species-preserving function (see also *PTN*).

4.3.2 Eidetic Plasticity

The changes brought about in the eidetic processes by a state transition or by training in abstract procedures do not require being remembered, i.e., being somehow stored as memory of the patterns of procedures. There may simply be an improved capability for conducting abstract reasoning, representing in turn the main functional characteristic of the new state reached by the eidetic structures after the transition. Therefore, it must be explained how the enhanced capacity is maintained forever, and how it may also undergo continuous improvement. It is apparent that we are confronted here with a problem offering two distinct, though obviously strictly related aspects. a) There is a qualitative difference between the eidetic operations before and after the state transition or before and after training, since as a result of these procedures a new eidetic operation —abstraction— becomes possible. b) The new property —the capability for theorizing— undergoes steadily progressive refinement with use, i.e., training effects are still obtainable upon reaching the new capacity after a transformation or after a suitable training period. The second situation could be interpreted as a quantitative effect, although the term refinement may also implicitly entail a qualitative improvement of the function. While it is conceivable that the mere maintenance of the newly emerged function —or capacity— is linked to memorization of new functional patterns (even if the actual mechanisms for stable, coded neural memory remain as yet unknown), the concept of refinement seems hardly applicable to the "stored state" of functional patterns.

Neuroscientists use the rather equivocal term plasticity for coping with such situations, due either to learning or to change in the structure (damage, growth) in which some rearrangement of the synaptic connections and of the related chemical transmitters are seen as a necessary stable basis to account for the changed or the newly acquired properties of the nervous system. Plasticity-based changes have come to be considered as mechanisms of memorization, not coded in genes or in other molecular sequences but in cell membrane synapses and chemical receptors and in patterns of connections, newly induced by as yet ill-defined stimuli inherent in the very different situations of learning and of structural changes. Neurochemists have enjoyed some success in isolating molecules endowed with receptor properties or with the capacity to stimulate a new formation of connections; yet, there seems to be no clear-cut evidence as to the mechanisms that define the distribution patterns of the receptors and the connections to be realized. However, the problem might also be considerably simpler. In fact, plasticity may be interpreted more restrictively as the process underlying the opening of unused or the facilitation of little used intercellular connections by reinforcing the production of chemical mediators at the synapses (perhaps, also directly inducing the formation of suitable receptors on the postsynaptic membranes). In this understanding of the term plasticity, no mechanism for defining a pattern of connections would be required since the new pattern would result from the opening to use of an already existing structural substrate, most probably through a local control of the chemical properties of the tissue (to be ascertained).

At this stage, an explanation put forward above for the eidetic operator seems simpler and more likely. The concept of eidetic operator was borrowed from the analogous concept in quantum mechanics to explain the procedures through which transformations of state are achieved in the eidetic structures. This operation needs not be highly organized. To induce the emergence of new properties in the complex maze of the neural structures, where each has a demonstrably different threshold, it is sufficient, at least in theory, to enhance or decrease their normal working level. This implies a simple, and simply produced, quantitative change, such as that accompanying every state transformation, either in the whole brain or only in part of the brain subsystems. In this case no procedure of memorization and probably also no plasticity seems to be necessarily involved since only a reinforcement or a facilitation of the normal state transition operating mechanisms would be called for —the process that occurs whenever attention to particular stimuli is the relevant factor. The determining factor is generally not the intensity of the attention-evoking, or clearly arousing stimulus but its information content. The latter entails reactions of fundamental importance for life (such as the typical case of the stirring, or of the voice of the newborn baby alerting the mother). Likewise, the strongly attention-catching eidetic problems may easily be assumed to play the role of very sensitive trigger for the transition operating mechanism, essential not merely for the eidetic function in general but specifically for the abstraction process, on the one hand, and for the so-called plasticity, on the other. The automatism of triggering inherent in these reactions can only stress the fundamental significance of the intellectual operations as man's characteristic life-preserving function.

4.4 Motions of the Ideas

In the preceding sections, we have seen the two paths of reasoning, analysis and theory, as proceeding, figuratively speaking, in two opposite directions, both because their goals are at the two extremes of an ideal unitary chain of thought and because the heuristic procedures involved in the two modes are not reversible. This is a strange and rather unique situation. When starting from a well-founded, comprehensive theory, one can proceed with the analyzing procedure in an entirely natural fashion, such as going "downhill," whereas the "uphill" way requires a total change of strategy. We might say that there are two different and completely separate routes. The belief in such an asymmetry led Einstein to claim the lack of logical links from analytical to theorizing thought. As a consequence, the latter should be considered as illogical in terms of the logic of the former.

The essential analytical procedure is the search for the way in which element-ideas are bound by relationships of clear cause-and-effect type —constituting the general case of reasoning in scientific research— or the recognition of necessary relationships among the element-ideas, such as the special case of consequentiality in pure mathematical reasoning. In order to avoid misconceptions about the true nature of this type of procedure, one should be clear about the meaning of cause-and-effect relationships, the basis for judgement about this, and the way in which the judgement is made. The analytical procedure of seeking and recognizing consequentiality (a less committed way of more generally indicating the class of relationships including the cause-effects ones) can easily be founded on the same comparative operations of congruence-coherence believed to be fundamental for all brain functions, autonomic as well as somatic. As has been stated repeatedly, the analytical procedure provides the sequence of comparisons constituting the basis of the logical or "scientific" reasoning, which is made of chains of congruent ideas and propositions. Quite to the contrary of the case in theoretical procedures, the search for cause-effect or, more generally, for consequentiality does not proceed by creating ever more abstract constructions out of basic or already grasped concept-ideas. Rather, it tends to split the large ensemble of elements of a proposition (a problem or a sentence) into smaller ensembles and thus to unravel the details of the machinery that holds them together, not of the principles governing their organization.

The typical reasoning of scientific research, in physics as well as in biology, goes into the actual structure of matter and describes the relationships among its elements and the forces determining its motion. These characteristics do not describe the basic function of eidetic processes, since these deal specifically with the formation of abstract objects, a process in which nothing is really immediately or already grasped because every object (either idea or thought) is formed through a congruence of elements that must be recognized before the "birth" of the object. Through analysis we can examine our own thoughts —precisely what parts of them have been transposed into structured language— by splitting the chains of propositions, by recognizing whatever form of consequentiality (cause-and-effect relationships or other) binds them, and by eventually disintegrating the abstract objects to form less abstract, simpler and ever more matter-related elements. Thus, also in eidetic activity analysis means the search for the particular, the downhill motion to the details. This going downhill implies that while unraveling the detailed machinery of the objects, on the one hand, one also discovers the finer relationships that rule the machinery at this more detailed level, on the other. (These relationships are the essence of working hypotheses or partial hypotheses which have the function of addressing the search in any one of the many dichotomizing paths met during scientific investigation.)

The arguments in Chapt. 1 about eidetic motion take on a new meaning here, for we seem to be dealing with a natural or quasinatural motion, although not properly in the exact sense of statistical mechanics, since it is not left to chaotic unpredictability. This motion is natural in the special sense that it is the brain's immediately spontaneous way of proceeding when engaged in problem-solving tasks; this does not yet imply a specific mechanism in this spontaneity. One might suspect that in these instances the eidetic function is, as it were, automatically guided along the path of thought offering less "resistance," the path of the more easily understandable relationships —those among the elements of matter— the path leading deeper into the details of the links within that complex of objects subjected to analysis. This leads to a conclusion with a clear philosophical meaning: the relationships that one discovers in science about natural and nonnatural objects are only those which the brain is able to understand. Although this may appear to underestimate the role of brain in investigative analysis, this is not actually the case. Every scientist engaged in investigation, particularly that of physical matters, knows well that almost every step calls for a decision, and that the true —i.e., not uniquely and not previously determined— decisions are always taken according to some general principle, to some belief that has matured within and is founded upon a more abstract vision of the problem, in other words a philosophy suggested by preknowledge. Analysis cannot represent anything but the discovery of the logical, natural path of development from these matured philosophical standpoints towards their natural, logical consequences. The motion is downhill, it is true, but it is almost never an easy one in practice since to be able to understand the logical relationships among new objects one must always recur to other principles, to other abstract relationships. Thought continually shifts back and forth from theory to analysis, dismantling the theory with the analytical procedure, using the result as preknowledge to refine the theory, and checking it again in the endless search for a unified picture of the system, the goal sought for —either consciously or unconsciously— by every scientist.

4.4.1 Brownian Eidetic Motion

Recalling the arguments in Chapt. 1, we can now understand natural or quasinatural motion also in the sense that each element-idea or each elementary proposition in thought may be subjected to minute transformations. These can be such that each one may be considered the "cause" of the immediately following one, which in turn is thus the "effect;" put more simply, in the transformation an order, constituting some type of consequentiality, is determined. Although the transformations of state —corresponding to changes of reference framework during eidetic activity— have been depicted as being performed by an eidetic operator, the question is still not really settled. The view in statistical mechanics' sense of a system made up of instantaneous "elementary" state changes occurring in its myriad variables (see *PTN* for the discussion about neural states) continues to haunt us. The natural motion of such a system or ensemble of such systems, in Gibbs' sense, has a very attractive, natural simplicity and the important advantage that it does not require additional structures or organizations to account for its own development. On the other hand, the ensuing disordered, casual motion of the system, typical of the statistical mechanics' view, seems to fit reasoning only on the basis of a kind of trial-and-error procedure because detecting cause-effect relationships or simple consequentiality would otherwise be completely impossible. Furthermore, nothing could be said about unfailing modalities for detecting cause-effect relationships if the resulting new element-idea or elementary proposition were the outcome of a different type of cause transformation (e.g., simply not minute ones) since in this case the "form" of the effect could well be conspicuously different from that of the cause.

Therefore, let us consider a bit more closely the hypothesis of a casually proceeding eidetic motion, a kind of Brownian motion of the elements of the ideas, or of the element-ideas in elementary thought propositions. This picture is no longer as preposterous as it appeared at first sight. We know that the number of inputs to each neuron and to each circuit of neurons of the central structures may amount to several hundred. In these conditions, we may assume that in terms of nerve impulses the activity of the neurons does not statistically change (i.e., it does not change in quantitative terms) in closely succeeding time intervals. However, we cannot likewise assume that the information contained in this activity also remains qualitatively the same since a statistical reasoning of this type cannot be applied to it. As a result, we should expect constantly appearing, if not regular or goal-directed, variations in the content of an idea or thought when the neuronal circuits embodying the ideas are engaged in a steady-state natural motion of statistical type. While we would be inclined to reject a priori the hypothesis of natural statistical motion for an essential and peculiar function of humans such as the formation of ideas and thought, closer examination shows that it could actually be seen as a mere mechanism of motion, i.e., for keeping the system running (albeit implying some weakening of Gibbs' constraints).

One of the principal questions discussed in *PTN* was the ability of the brain to maintain a constancy of functions despite the variability and instability of individual neuronal activity. Precisely this characteristic suggests that brain operations are based upon large ensembles of such individual chaotic motions. The necessary relaxation of Gibbs' conditions (large number of similar and not interacting systems) consists of the fact that the neurons of the central structures are largely interconnected and thus interact with each other to an extent hitherto unforeseen. As a consequence of their interconnectvity, the casual motion of each neuron and therefore of each circuit affects the motion of many others, in such a way that: a) each neuron and each circuit bestows on the others some of its own instability, but b) actually no neuron or circuit is completely free to follow its own chaotic motion (in *PTN* the resulting motion is likened to that of a deformable body). Consequently, in the motion of such an ensemble of single neurons, as well as of eidetic subsystems, there should always be a relatively high correlation among the elements and therefore possibly also in congruence-coherence regarding their information content.

4.4.2 Ongoing Eidetic Motion

The proposed statistical mechanics' view of chaotic motion, although feasible, is not a prerequisite. If we accept that the neurons involved in eidetic functions (such as building ideas and propositions) are engaged in a comparable degree of ongoing activity and are endowed with the same high capacity for interaction, we can conceive of the intrinsic meaning of eidetic objects undergoing progressive changes, independently of every statistical mechanics' consideration. We also judge the changes as logic because there is a necessary correlation between the closely successive states of motion, and because this type of motion still represents the natural way of working of the brain, or perhaps the only way of "spontaneously" working (again, because it is probably the one following the path of least resistance). However, the previous conclusion that the relationships which we recognize among the objects of thought are the only ones that we can understand is at first sight likely to be highly disturbing. It emphasizes the hidden character of relativity inherent in all our brain judgements, sharply contrasting with the view of positivistic science that those which we detect are the true (and sole) relationships. Accepting this type of philosophical relativity —actually not a measurable degree of uncertainty— in eidetic processes amounts to maintaining that science is fictitious; it is a construction in which we are unable to detect weaknesses whenever their origin does not lie in the presumed links among the objects but in mind itself, its methods and its standards of measurement and comparison. In the "Appendix" to *PTN* we also raise the question of these standards, but we see here that the answers are not simply in the domain of neurophysiology, psychology, or philosophy alone. The answers cannot be provided by the usual analytical methods of science because these unavoidably lead to the contingent and variable, i.e., toward the relative in the relationships among the elements of an ever-moving nature.

We have said that the type of motion in a deformable body (itself constituting a nonchaotic motion of its parts) explains how reasoning may proceed from idea to idea through a series of minute changes —a piecemeal development of thought. There is no danger that the ideas linked one another in this way can be judged noncongruent and noncoherent (illogical) because the logicality of their connection is ensured by the fact that at the very moment of their appearance we see them as new rings added to a same chain, and that we cannot see them otherwise. In fact, there must always be a high degree of congruence-coherence among these ideas or statements because each one results from a minor transformation of another. (This is somewhat as with the various meanings listed for a single entry in a dictionary.) Moreover, no operator is required because the minute-step transformation could be contained implicitly in the concept of ongoing, spontaneous, or natural motion-activity of the neurons (although it cannot be excluded that an operator is required for ongoing activity as well). Thus, the principle of natural motion, either in the statistical mechanics' sense or not, which at first sight seems to create chaotic motion in an ensemble of independently moving elements or subsystems, actually gives rise to a sequence of steps that are connected logically, i.e., largely congruent-coherent with each other (though probably not goal ordered) provided the degrees of freedom of the elements or subsystems are effectively reduced by their anatomical and chemical cross-connections. Only in the unreal, extreme case of a network of connections chaotically determined is the uncertainty of the resulting motion strongly increased since the elements for constructing the ideas and the thought could be randomly dispersed in all possible directions, thereby reducing the probability of creating new, congruent objects.

However, on second thought, this may seem to involve contradiction. It has just been said that a sequence of ideas is judged to be logically connected for the very reason that each idea is a minor development of a preceding one. However, also in the case of a chaotic connection network each idea represents a transformation of another one —if we still maintain that natural motion proceeds by small steps. Therefore, we would judge each appearing idea as a logic consequence of or logically connected with the previous one although made up of elements that are not congruent-coherent because we would continue to see them as serially successive rings of the same chain of thought. There seems no escape from this type of ambiguity inherent in the stepwise transformation procedure, either randomized or not. This pitfall can be avoided, however. Reasoning does not always link two ideas of whatever sort into some logical connection; chaotic or simply ongoing eidetic motion constitutes nothing but the running mechanism, and every transformation is subject to a judgement on logicality based on comparison of the intrinsic content of every idea with the preformed scheme of the thought into which it must fit. In other words, the "rough draft" (of course, not the full development) of the thought —for instance, some simple sequence of congruent-coherent significances— is present as a starting point, as a guideline linking the premises, which can be completely and precisely defined, with a final that might be more or less clearly "seen in advance." (The ideas may be viewed as spikes hammered by a mountain climber into the rock and to which the rope of reasoning is fastened step by step, after the path from one to the next has been selected in advance by comparing the possible alternative routes.)

Let us examine more closely the alternative choice aspect of idea and statement transformations. It is apparent that at every minor transformation there may be a number of choices represented by different shades that the idea or statement may take by changing only one or the other of the elements composing it. Potentially, each choice gives rise in turn to another line of transformations, i.e., the continuously dichotomizing paths may lead to progressively divergent endpoints, to more differentiated ideas and statements. An analogous situation is considered in *PTN* in relation to state transitions. These were conceived as gradual (but more or less fast) transitions from one physiological state to another through a number of intermediate microstates. However, the microstates cannot be precisely defined physiologically because the variables defining a state are all the variables of the system, always in an extremely large number, and, in principle, to produce a new microstate it is sufficient to change only one variable. In the case of transformations of ideas and statements, however, the number of elements involved is certainly much smaller, so that also small steps could produce noticeable changes of an idea's content that are recognizable because they are likely to involve a turn in the line of reasoning.

4.5 Field Ideas

Thus, ideas gradually change their information content and give rise to trains of other, somewhat similar but already clearly differentiable ideas. Of course, reasoning is not made of these trains of ideas. Reasoning assembles different ideas following a roughly predrawn scheme of thought (see above) of which only the starting and endpoints may be clear at first. However, the heuristic procedure of idea transformation, on the one hand, and of thought development, on the other, need not be different. In either case we are confronted with the development of information contents through gradual changes of the constituting elements, so that at every step we can recognize congruence-coherence; in fact, the procedure comes to a halt as soon as the congruence-coherence is lacking, that is, as soon as mind is unable to detect the logical "contact" between what precedes and what follows.

Let us look the idea with its many elements, together with the subtle changes that the same idea may undergo and the many dichotomizing paths developing from the idea and its transformations. The picture that one obtains is that of a "core" of the idea's content surrounded by a periphery containing shades of the original meaning (i.e. slightly modified ideas) which still bear the mark of their origin since the source of their derivation is apparent. The whole complex can still be identified by its central core of content, which extends its "influence" throughout a region of the space of the ideas until regions are reached where the kinship with that originating core is no longer recognizable or has become ambiguous. Considered in this way, the idea core becomes a field-idea, an abstract structure because of its very abstract nature, with many connotations of a (circumscribed) continuum. The field concept needs more explanation when applied, as here, to an abstract object such as the idea. The originating idea, from which the transformed ideas originate, and the transformed ideas constitute a continuum, inasmuch as in every region of the space covered by the ensemble of these objects (the ideas) there are elements of the original one (according to the physical definition of a continuum).

Let us see, then, whether the field idea has other properties pertaining to the continuum and to field concepts. We can begin by asking whether the field idea constitutes a homogeneous continuum, i.e., whether the field is homogeneous; for this, the elements of the continuum must be evenly distributed. This property cannot be attributed to every field idea because homogeneity would depend on the significance of the ideas and on the possibility that every element of them gives rise to evenly distributed dichotomizing paths. In fact, we should expect some of the ideas derived from the core to give rise more fruitfully to a larger number of alternative paths than others, suggesting that the field must necessarily be nonhomogeneous. On the other hand, we may safely assume that the field idea will be anisotropic, since the properties, i.e., the qualitative aspects, in no region of the field can be the same; otherwise the ideas of the field could not be distinguished from one another.

With the existence of a field there must be "eidetic gradients" originating from the idea core. We can consider the capacity of originating new ideas, a motion developing along dichotomized paths, as the measure of the "strength" of a field of force originated from the core. The density of distribution of the paths outlines regions in the field of eidetic force (in the field idea), with the lines of force showing the preferred directions along which the idea core develops (as we shall see below, the common content of the term development does not portray the true characteristics of the field idea's nature). The farther from the core, the smaller is the number of alternative paths that each transformed idea can produce because the number of elements left for subsequent transformations is smaller. However, at this point it is awkward to make hypotheses about more precise connotations of the field gradient and of the correlated force involved in the field idea. Nonetheless, if the concept of the field idea with all its implications is accepted, the arguments examined above in relation to the mechanisms of idea transformation may help us get nearer to the point.

The choice is always between natural and operator-effected motions. As regards natural motion within the field idea we can now take the existence of a gradient that by itself explains motion. Of course, "natural" no longer would signify a random motion in a statistical mechanics' sense but the natural way of moving, provided by the energy gradients in the field and guided along the lines of the corresponding force. These are expected in turn to correspond to the regions where "resistance" is less, in our terms where congruence-coherence between the ideas before and after transformation is higher, i.e., where the difference is smaller. Despite the absence of a reasonably clear-cut definition of its nature, the gradient should be seen as playing a role equivalent to some dynamic, thermodynamic, or electromagnetic parameter. The motion throughout the field idea can always be represented in Lagrange's generalized system of coordinates (the method suggested in *PTN* for all motions within the central nervous system). This issue is undoubtedly preformed in much more abstract terms, since we are no longer dealing with the force of the nerve impulse trains as such (this being the general case for every brain activity) but rather with an eidetic force itself, or with the gradient of an eidetic energy in whatever terms, physical or abstract, we may calculate it. However, we can put this much more precisely by maintaining that we are dealing with a gradient of "logic," and that the lines of force are the lines along which the "logic" of the transformation encounters the least resistance (see also Chap. 10).

Concerning this last point, some further considerations are necessary at the level of neurophysiological parameters. The force of the impulse trains is held to be measurable through the effects produced, i.e., the change in momentum of the activity of the neurons submitted to nerve impulse bombardment. Therefore, the effect depends, as usual, on two orders of variables, the characteristics of the impulse train (the force) and the resistance of the neuron membranes (a function of the preexisting activity, together with inhibition and a number of other membrane factors). Given parity of conditions, we should expect similar effects; otherwise it would be useless continuing with the procedure. While in the case of somatic and of autonomic functions we can fairly reasonably expect similarity of effects in similar conditions of neuronal activity, we cannot be as sure when dealing with eidetic development. The reason for this is that the recurrent motion from one idea to another should, to obtain similarity of effects (or of followed paths), involve similarity of distribution of activity in the eidetic circuits concerned, if not in different subjects at least in the same subject on different occasions. However, this does not seem very likely either in different subjects or in the same subject at different times and in different conditions. The development of ideas does not appear to be so mechanically determined as a mere eidetic-energy gradient would entail for a natural motion, and therefore other factors must be considered. If we maintain that the same idea must be represented in each different situation by an immutable, though complex pattern of activity in an ensemble of circuits, i.e., a fixed constellation of distribution of activity in the neurons constituting that subsystem, it follows that the energy gradient, the logic's strength itself (i.e., the force), must be endowed with the possibility of varying while leaving the motion between the ideas unaffected. Reversing the reasoning, the energy gradients should not be inherent in and therefore not directly representable by the strength of the nerve impulse discharges (which may easily vary according to the "state" of the system in different conditions) or at least not representable by the usual elements of neurophysiological activity but would depend exclusively on congruence-coherence, the true "force" linking the ideas, a force that may vary only according to different contexts of thought.

4.5.1 Operator-Dependent Motion Through Field Ideas

We shall now examine the possibility that the motion through the field idea is an operator-dependent process. Strictly speaking, after what we have said up to now about natural motion within a field, the presence of an eidetic operator within a field idea would appear altogether unnecessary. The existence of an eidetic energy gradient from core to periphery of the field, assuming the concept to be sufficiently clear-cut in its implications, may account for a motion through the different shades of the idea core, until the border of the field is reached. What might be found at this border is only matter of speculation, but we shall try to define better what is meant by border. We may assume that a field idea, like any other field, vanishes at the point at which the strength of the field is nil, i.e., when an influence of the core is no longer detectable. In this case one can see the progressive transformations as having exhausted the stock of possible changes of elements related to the core content of the idea. If this picture is accepted, at least tentatively, the field, although extending the dimensions of the idea core to include a variety of shades, still retains its intrinsic monadlike content. Perhaps, the monad is reducible further (if we may say so) but is not expandable, a characteristic sufficient for isolating it from other, similarly monadlike abstract structures. Upon closer examination, however, the isolation also seems not to be very strict. The elements, constituting in their ensemble the idea, are only exceptionally exclusive of a given idea; generally they are common to several. This means that different field ideas may merge at one or more points (i.e., in correspondence to the common elements) even before their respective borderlines are reached. The consequence is that we may be able to pass, more or less inadvertently, from one field idea to another. With this observation, what we have called the monadlike structure of the field idea becomes ambiguous, or vanishes altogether, and a connected network of such field ideas replaces it.

One might ask whether the path may be reversed, i.e., moving from the field border of shade ideas toward the idea core. For this reversed direction no suitable energy-gradient would be available because the direction of this is provided only from core to periphery. Therefore, if we are to hypothesize such a reversed direction, we must recur to operator-induced transformations as the only type that could provide the necessary energy. Some of the aspects of this situation are reminiscent of the discussion about the asymmetry of analyzing and theorizing procedures. These use the same elements, but the path of reasoning proceeds in the opposite directions. We concluded above that analyzing and theorizing use two different heuristic procedures, the downhill motion (analysis) being the simplest and therefore easiest and most natural way of progression whereas uphill motion (theorizing) consists of abstraction procedures and requires steady training to be developed and eventually also to become natural. These are therefore two different procedures and thus two different modes of thought development. Likewise, the back-and-forth motion through the field idea would seem to require on the same grounds different procedures. In particular, an operator-induced transformation would seem to be necessary for moving from periphery to core, which would retain the overall significance of a return to the origin, since it corresponds to regaining the more general and more comprehensive content, that of the core, gradually lost in the transformation from core to periphery (physically translated, this would correspond to the effect of a restoring force).

In Sect. 1.8.2, the interaction between operator and state was said to induce a modification in the operator itself, or the appearance of a different operator. (In such a process, on the one hand, the performance of the system was progressively refined while, on the other, the necessity of postulating a memory trace of the whole setup —operator and state— was avoided.) The interaction between operator and ideas, if it takes place at all, brings about a change of the operator at every step of transformation, which would ensure the attainment of a progressively more comprehensive content, perhaps not feasible with an invariant operator. The transitions from core to periphery can thus occur simply along the lines of force peculiar to each field idea, under the control of the eidetic-energy gradient, itself characteristic of the meaning of the particular field idea, but not necessarily in its presently detectable neurophysiological correlates (at least as quantitatively assessed). The transition from periphery to core seems necessarily dependent on will, as is the theorizing thought, a determined operation with a definite goal to reach (the more comprehensive content), but a goal with form and precision of content unknowable until they are reached. Therefore, every exploration of a field idea, for instance, within a particular context of thought, is not in itself a motion along an eidetic-energy gradient but actually a counter-gradient motion; it should thus always be looked upon as a will-directed thought, i.e., a motion directed by another form of energy (see Sect. 5.4). However, the will-directed operator may also determine the downhill motion within the field idea since the existence of the gradients from core to periphery and the factors determing the lines of force still constitute theoretical matter and may more simply represent the facilitation of voluntary operations along some directions (see also Sect. 5.1 for further questions concerning the assumed motion within field ideas).

I think it is clear that the mechanisms discussed so far —random microtransformations, unit operators, development of field ideas, downhill motion along energy gradients, and lines of force— do not require a judgement on logicality at every step. Regarding these procedures congruence-coherence is already implicit in the nature of the transformations or in the field concept. The judgement would become necessary and unavoidable only if the procedures led to a final idea substantially different from the initial one, but this should be excluded in the case of the microstep transitions on the basis of the common matrix of the ideas stemming from an idea core. However, whenever two fields are said to be merging, i.e., when they come into contact through commonly shared elements, the judgement of congruence is always called for because the mere sharing of some elements may mean merely superficial, formal likeness and is not necessarily an index of substantive kinship. In fact, the two ideas may belong to different regions of the phase space of the ideas, so that their essential kinship can be recognized only after the core of the two fields has been reached with an uphill motion, and after an abstraction process has possibly assessed the likeness of their intrinsic contents at a more comprehensive level. Such a judgement always considers the essential aspect of the things, disregarding differences that may be considerable in the superficial form of the abstract objects and in the conditions of their appearance.

5 Logic, Will, and Eidetic Function

The physical mechanisms of eidetic drive examined in Chap. 4, such as the change in characteristics of eidetic (and brain) space time, the field properties and the gradients of eidetic energy, do not avert the necessity of other, (as yet) not physically identifiable drives of motion, such as the eidetic operators. The existence of physical mechanisms with self-sufficient operative drive capabilities, however, does not exclude that they may become operative through other factors (or principles) which to be effective need to assume in the brain the form of a physical mechanism. In other words, those that we have identified as physical drives may be only the physical expression, i.e,. the material form, of more abstract principles.

In the present chapter we consider two types of nonphysical drive. The first is inherent in the relationships among events, not in their physical expressions but in the logic linking them to one another. We already have suggested that logic may play a functional role in the theorizing mode of thought by constituting the means for seeking a missing logicality in reasoning. While we have not yet attempted to define what we call logic, in the present context we take logic to be embodied in the eidetic circuits and in a way that it may exert a force or be a source of eidetic energy. (We shall not discuss now the problems connected to logic; these are considered in Part V)

The second type of nonphysical eidetic drive needing to use physical structures is will. We have suggested already that will can be seen as a matter-of-fact capability of mind for influencing mental activity (as well as brain activity in general) toward selected paths and predominating over every other drive. As we shall see, logic is a matter of the relationships among objects considered by mind and is also perhaps a problem of the structure of mind-brain. Will, on the other hand, is deeply rooted both in the nature of mind and in the psyche. We might say in fact that will is the principal manifestation of the human psyche and perhaps the most characteristic property of human brain nature.

5.1 Cause-Effect Relationships

In discussing motion through the field idea, we approached the problem of determining the logicality of transformations of the idea content. The conclusion was that if the changes are produced by a downhill natural motion along gradients of eidetic energy, there would be no need, in principle, for such a judgement on logicality, since the transformations involve the various elements contributing to build an idea only one at a time. (Of course, the problem ceases to exist if it is assumed, as in Sect. 4.5, that the essence of the eidetic energy gradient is actually the logicality inherent in the graduality of transition from one idea to another.) A judgement on logicality would also not be necessary in the case of an eidetic operator transforming the ideas as a "contact" or "unit" operator would do, i.e., by successive, minute, unitary steps. Since the latter instance, however, would represent a rather particular type of operator-dependent transformations, it would not generally obviate a judgement on logicality in cases in which an operator is involved. This point is fundamental in connection with the voluntary character of almost all eidetic processes leading to thought formation. The will-guided, goal-conscious effort to develop reasoning in a clearly defined direction is certainly not restricted to unitary changes of an idea's content; on the contrary, thought is generally expected to take "leaps" into other regions of the idea space, looking for different ideas, analogies, etc. Thus, a judgement based on the congruence-coherence measurements among objects being compared seems unavoidable as the sole procedure ensuring logicality throughout the developments of thought. We must now analyze the different forms which the judgement on logicality can take.

At first sight, the relationships or the logical links which relate the elements or ideas of thought among themselves are disguised under very different aspects in different cases, but we shall see that these can be reduced to a small number, if not in fact unified altogether. Taking as a model the more common physical situations, we shall consider three cases. The first is that of *post hoc ergo propter hoc*, in which a mere temporal, not natural or logical, order in the appearance of things is taken as evidence of causality. The second is a cause-effect relationship. The problem of defining cause-effect relationships has given rise to endless philosophical questions. Here we try to avoid the question by defining in advance the limits of the relationships that we want to investigate. A relationship of cause and effect is taken here in the common sense of an action or force necessarily preceding another, the latter being considered as a result of the former. Here, too, a temporal feature is involved, but time is not the essential factor; the necessity or inavoidability of the relationships between the factors is on the foreground. The third relationship is that of logical precedence (without temporal precedence), a kind of hierarchical ordering of the ideas. This is the most common situation in scientific thought and is that which is generally called rational, or logical reasoning.

In the eidetic domain, the mere temporal precedence of concept-ideas with respect to others is of very little importance. The ideas are abstract structures representing, with a progressively refined precision, the essence of some object, either material or itself immaterial from the start. As such, the nearer the idea is to an essence, the more it enjoys a timelessnes (see below); any construction consisting of such objects has no relationship to time. Although they can be ordered in terms of origin (one idea leading to others), the nature of such an order in no way resembles that of a chain of physical events having an origin recognizable in some prior event and following one another in a necessary time sequence up to a final event. The picture of ideas as being in series, or in a train suits the concept of field idea, according to which trains of ideas originate from the idea core, regardless of whether a unit operator is conducting the transformations, or transformations follow each other along the eidetic energy gradient of the field. However, in neither instance do the successive transformations depend on the idea core (the origin) or on time. One might be tempted erroneously to envisage an analogy with the inertial motion of a body, which maintains a constant velocity as long as no gravity (accelerating) forces are present. Here, the successive space coordinates (the positions) of the body are a function of the time from the beginning of movement and the relationship to the force of the originating event, while the first time derivative of the movement is constant. This argument has no eidetic counterpart since, in contrast to the physical analogy, the coordinates of every shade idea in the field idea are timeless, corresponding to its immutable position in the field; this is maintained by a time-independent eidetic-energy gradient, and the transition from one to another idea may also be considered a truly time-independent phenomenon, regardless of whether along the downhill gradient of the field or guided by a constant-motion, moving operator. (This picture is considered more thoroughly below.) We are forced to conclude, therefore, that time relationships do not enter into the judgement on logicality between ideas, although, as we shall see below, time may still represent a fundamental factor in eidetic problems, as in the physical ones.

Comparing eidetic motion to that of a projectile shot from a firearm may better serve to clarify the cause-effect relationship: a preceding action, the ignition of the powder, causes the motion of the bullet, the effect. If no other forces are present after the initial acceleration, the bullet's motion is rectilinear and time independent. Here, the judgement on cause-effect relationship at any moment of motion is no longer required. While we could represent the inertial motion as a chain of subsequent, separate, unitary events, each in some sense caused by the preceding one, i.e., as its transformation, this would simply introduce an artificial time dependence of forces where there actually is none. A physical law predicts the position of the projectile, given its mass and initial velocity, from the beginning of flight, and this is the only time dependence to be considered. The law governing the effects of the initial action alone ensures the rationality (if we prefer, the logicality or the congruence) of motion at any moment. A deviation from the rectilinear motion may be introduced by intervening causes, such as gravitational attraction, but these too obey a law in defining the future course of the projectile. Therefore, whenever we find in eidetic motion that type of predictability as in the Newtonian laws of motion, we have an intrinsic, implicit judgement on logicality and congruence.

On the other hand, however, the analogy with the projectile's motion is not appropriate for understanding what occurs during eidetic development regarding the ideas within a field. Each idea in a field, considered alone, has an existence, but the idea core still keeps under its influence all the shade ideas, in a way that although not implying any time relationship between them defines a rather clear-cut, logical dependency of every idea on the immediately preceding one up to the idea core. If we accept the picture of the transformation of the idea (of the field) into another, operated automatically by the force provided by the eidetic gradient of the field or by an eidetic operator, we assume the same type of cause-effect relationships between the ideas as exist in the inertial, time independent phase of motion in the bullet's flight. We can adopt the same arguments to explain the interference of other forces (operators) able to condition eidetic motion. Regarding field idea, however, the arguments concerning accelerating forces are completely out of place. The concept of the field idea entails the simultaneous presence of all the ideas belonging to it, already ordered according to a principle of logic which recognizes an origin in the idea core and starting from the latter connects the other ideas one by one. No event needs to take place, not even recognizing the logicality of individual links, since these are already inherent in the logic of the field organization. Thus, when introduced into reasoning, the ideas of the field are all simultaneously present —the idea core as well as the associated shade ideas— and no particular idea needs to be discovered by a motion started from the core.

Obviously, there would be no objection in principle to the existence of an eidetic gradient between ideas not belonging to the same or to any field at all; nor as just stated, would there be an objection either to the presence of operators providing forces other than the eidetic gradient. We have already based our arguments about theorizing motion on the existence of operators generating motion neither along a natural gradient or along the natural direction in the development of the ideas. The question is whether we can look upon the relationships among ideas in terms of cause and effect, as in the physical instances presented above. How can we say that one idea "causes" another? We seem in danger of being dragged into the philosophical discussion that we wanted to avoid at the beginning. As in *PTN*, we shall maintain the position that the brain is a physical object and operates only according to the ideas physically embodied in its neuronal circuits. This leaves to physicists and philosophers the debate about definitions of cause and effect. We shall further assume that cause-effect relationships between ideas exist only if these relationships can take the form of physical laws, such as those of motion, since we have consistently maintained that the activity of the brain is to be considered a type of motion. From this point of view, relationships of cause and effect are always based upon a physical law or to an analogous principle involving the same type of relationships between its elements. This standpoint rests on firm bases. The very formulation of a physical law is the transcription in symbolic form of a reasoning (i.e., a sequence of ideas translated into symbols) setting the elements into relationship with one another to represent the mathematically analyzable aspect of that law. Whether this can be interpreted as representing cause and ensuing effect remains debatable (the point we have sought to avoid). Physicists claim that there is nothing in the physical laws, such as those of Newton, that may expressly indicates a cause and an effect, due to general symmetry of formulation and the absence of explicit reference to time. Other positions are taken by philosophers regarding the concept of causality.

For our purposes, of paramount importance is the existence of an isomorphic correspondence between the ideas as elements of reasoning and the elements of a physical law because a judgement on logicality or on congruence-coherence among ideas is implicit in the fact that the ideas can be so ordered as to express a law. This correspondence represents the third type of relationship considered above i.e., that of logicality based upon the necessary hierarchical antecedence of one idea with respect to another. As we said, this type of relationship between ideas forms the common basis of reasoning in the scientific domain. Thus, the idea B presupposes the idea A, and if A does not exist, neither does B. Every proposition in science is based on this pattern of thought or recognizes it ultimately as its origin, in mathematics, physics, chemistry, and biology, albeit with different degrees of dependency by B on A. (The relationship of logical antecedence is essentially of the same form as the cause-effect relationship, a necessary antecedent and its result.)

One should stress in eidetic operations that the necessary antecedence is only logical, not temporal, and that the principle of logical antecedence is the only one that allows us to reach a deeper understanding of the things of nonmaterial nature. In fact, without this principle our reasoning would remain confined within the realm of physical events where there always is temporal and logical antecedence and consequence, but where the temporal aspect strikes the observer and may obscure the logical aspect. The relative inconsistency of the temporal factor is highlighted when one considers that, as there can be no effect without cause, there can also be no cause, in strict sense, if there is no effect; this becomes apparent in practice when a cause is prevented from acting —in this situation both cause and effect remain but potentials. As soon as the cause is permitted to act, however, the effect is simultaneously present; as soon as we stop an effect (for instance, the fall of a stone), the cause effectively disappears too, in the sense that it remains a potential. The inadequacy of common verbal communication is eliminated when cause-effect relationships are symbolized in linear equations; here, each element is symbolized and suitably shuffled around, with the equation retaining its fundamental logic-based content. The simultaneous presence of cause and effect in the symbolized physical law is also found in the embodiment of the idea in feedback circuits of the implicit type (see Sect. 1.7 and Chap. 8), as far as the idea may be represented symbolically as a physical law.

5.2 Hidden Cause-Effect Relationships

We look now at general principles to explain logical links between concept-ideas that escape the logic of mathematical formulation. We must point out, firs, that we are not dealing with instances in which idea B is a direct transformation of idea A; this involves merely a judgement on congruence-coherence, implicit in the operation of transformation, not a relationhip of dependence. Logical dependence obtains only when A and B are combined in a particular type of proposition —not a proposition directly understandable as a physical law, however, which is covered by the cause-effect relationships examined in the preceding section. Thus, we are not concerned with physical facts in themselves but with necessary logical connections between A and B that are not (or not immediately) evident and thus not recognizable.

What we are dealing with, then, are instances in which the proposition A-B appears illogical or contrary to "sound judgement." Because a logical coupling of the elements A and B does not on the surface suggest itself, the logicality of the link between them would not be sought. That a link does, however, exist may become evident through an improbable or unexpected event. Disregarding the trivial occurrence of a fault in judging the logical consequentiality, the obvious conclusion can only be that elements must have been overlooked, and that the statement of premises was therefore not complete. However, such incompleteness cannot concern elements of the same nature or order as those belonging to the ideas under examination because this would be, again, a trivial case of misjudgement about an otherwise easily recognizable logicality. Rather, the missing elements must pertain to other regions of knowledge, and their logical position in this context must be submitted to judgement. Thus, the reasoning may also seem in the beginning to progress in a completely unforeseen and perhaps apparently illogical direction, but this is only because mind has become aware that other elements also need to be considered, and having done this, the situation may move to a different but still appropriate (logical) outcome. It is obvious that mind's judgement, the successive steps of thought, and of course the resulting behavior are always expected to comply with each other, but they may do so exclusively on the basis of the more appropriate choice between alternatives. The latter, on the other hand, cannot be evident until all the relevant elements themselves have also become evident.

Let us consider an example discussed in the "Appendix" to *PTN*. The concept-idea "thirsty man" is logically linked to that of "drinking" and precedes it —let us say it is the cause of it. It may well be the case that notwithstanding the man's thirstiness (A) drinking (B) does not result, rather, non drinking (C). While the connection A-B is logical and can be viewed as a cause-effect relationship in elementary reasoning, the connection A-C cannot be interpreted in any way as the outcome of logical reasoning. It is obvious that a cause of non drinking (better, an element justifying it) must exist if we maintain that non drinking was a choice, i.e., an act with a rational basis. There may be several of these elements, for instance, social, ethical, religious principles, or affective and emotive drives. If this cause (or element, condition, hidden variable) is not brought to light, the absence of B lacks logical explanation since non-B, that is, C, should be produced only by non-A. The relationship between cause A and effect B, determined by and obeying a "physiological" law, is replaced here with the combination A-C, no longer a cause-effect relationship since C is not produced by (or is in any way dependent on) A but by the simultaneous presence of A and of one or more hidden variables. To represent the situation better we should say that the hidden variables have transformed the cause A into A* (as if a "complex" A) i.e., thirsty man possessing principles or motivations that logically explain the relation A*-C. We have replaced the simple physiological law with a complex but still physiological law (since these motivations founded upon ethical, social, and religious principles, together with emotive and affective elements, are for man physiological factors of judgement).

Logical causality is merely a colloquial way of expressing a relationship between (logically antecedent) conditions and (logically ensuing) conclusions, the basic concept being that the concerned conclusions can be drawn only if the suitably corresponding conditions are present.The conditions can be symbolized, and seen as coercive enforcing principles, or plain forces, regardless of whether they refer to physiological events of body or to abstract constructions of mind or even ethical principles. Even more than for physical laws, however, logical antecedence of the conditions does not imply any specific temporal relationship even if in many instances a precise temporal reference may exist. On the contrary, the logic inherent in such laws and in our previous reasonings prescribes that the conditions and the logical consequences (the conclusion) must be simultaneously present. The logical conclusion does not refer to temporally anteceding conditions but to the conditions present at the very moment —literally, the present— since obviously what was temporaly anteceding could very well be changed in the present. (Also the law of motion in the gravitational field, to be held as a reliable predictor of body motion in a future time, requires that the value of its time-bound changes at every moment are defined by the local instantaneous value of the mass-energy generated gravitational forces.)

Physical, biological, philosophical, ethical, religious, and social laws can initially be placed on the same level. The preceding example of a strictly biological law (the cause "thirst" the effect "drinking") can be seen as being as coercive a physical law. This is particularly evident in cause-and-effect relationships between neuronal events in anatomical structures, which are almost as rigidly determined as reflex arcs (a situation corresponding to the motion of a body with very few degrees of freedom, as constrained on rails or between walls). However, up to now such a coercive character is not generally recognized in broadly philosophical laws —or, conversely, these principles are not seen as having the status of a law. This means that our argument require the addition of some more arguments. First, there is a problem of semantics. I have consistently used the term law for nonphysical and not strictly biological principles as coercive elements of thought and behavior. However, most neuroscientists consider these broadly philosophical principles (with the exception of those intruding into the psychological area) as simply providing generic prescriptions or directions, and up to now these have not been considered as physiological elements of judgement, even less as "forces" actively intervening to modify an action otherwise dependent upon a biological law.

Let us return to a contention made in Sect. 2.3. The so-called rational —purely scientific and mathematical— logic is the basis of reasoning that consists of an unavoidable sequence of steps, succeeding each other in terms of the laws of consequentiality inherent in the nature of the object of thought. As already stated, the choices allowed in this type of reasoning are extremely limited or nonexistent, particularly when purely mathematical problems are concerned. Once started, the reasoning may proceed in a straightforward way —as an inertial motion in space without gravitational forces. Nothing is able to divert the course of reasoning as long as it remains within the boundaries of scientific or logico-mathematical objects. A "complex" logic, on the other hand, drives thought when the "simple," unavoidable relationships inherent in sets of general scientific (including mathematical) or matter-related elements clashes with other sets of elements, namely, the broadly philosophical ones which introduce a disturbance in the straightforward development of the former. Thus, the essence of these broadly philosophical elements is also "force", making the reasoning of complex logic a system which also includes mathematical-scientific reasoning, and the latter is seen in its pure expression (in the limit case) of "disturbing force null." From this it follows that the mere existence of a system of relationships, regardless of how complex, implies some degree of unavoidability of the consequences dependent on the intensity of the intervening forces. A sequence of reasoning steps may ensue that is not always, and not fully, predictable, and that possibly appears as "illogical" thought only if one ignores the nature of these new elements, i.e., of the force inherent in them. The analogy with the gravitational situation is emphasized by the fact that these broadly philosophical forces may assume different degrees of intensity within a wide field of action comprising both eidetic and noneidetic functions, thereby also causing differentiated transformations of the reference framework, i.e., differentiated changes in the local metrics in different regions —thus, not intervening simply as new elements of judgement for developing eidetic and for driving behavior.

5.3 Principle Ideas

The "effect" exerted by broadly philosophical principles upon thought development id closely analogous with the effect of a field such as the gravitational one; these principles also constitute fields, what we may call "principle idea" fields. The ideas representing principles such as those of philosophical, ethical, social, religious, or political content themselves appear as huge field ideas of a very complex type, costituted by ensembles of field ideas; this fact is so important that more comment is needed on this. The structure of a field idea is rather simple (see Sect. 4.5). An idea core which gives origin to other ideas strictly connected to by their intrinsic meaning and by their common origin; each may be thought of as deriving from another through changing one or a few of the constituent elements. The transition from one an idea to another was assumed to occur along an eidetic gradient progressively differentiating the shade ideas from the core until the traces of the common origin are altogether lost. Of course, the extension of the fields may differ considerably depending on the richness in meaning that may emerge from the idea core through the progressive transformation of its constellation of elements. Those that we will call principle ideas have a field structure much more complex than simple field ideas. First of all, it is very difficult to outline a unique idea core from which we may derive all the ideas of the field. Certainly, the core cannot be seen simply as ethics, philosophy, religion, or society. These terms are not sufficient to outline a unique concept. Rather, they indicate a whole region, an ensemble of concepts, themselves representing other ensembles; examples in these regions include the concepts of justice, good, evil, time, essence, intellect, conscience. If we push further the analogy with cosmic objects exerting gravitation, we could say that the simple field ideas form large ensembles corresponding to concepts —as the cosmic bodies, stars, and planets do by forming large ensembles, the galaxies. Concepts in turn form larger ensembles constituting eidetic regions, such as as ethics, philosophy, society, and religion, as the galaxies constitute larger aggregates, the galactic clusters, characterized by their coherent motion in cosmic space.

This may seem merely a fanciful image, but it is actually much more, for a fundamental quality in it makes of the comparison between the two situations a more meaningful analogy. Clusters of galaxies are astronomic unities, made up of innumerable elements held together by the combined forces of motion and gravitation over cosmic spaces in a coherent motion, i.e., rather fixed reciprocal relationships and comparable velocities;. in this way, each cluster is a reference framework common to all its elements. A principle idea is like a galactic cluster: it holds together large ensembles of concepts belonging to that principle, and each of these sub ensembles in turn holds together innumerable field ideas and ideas. Likewise, a principle idea constitutes by itself a reference framework in which the sub ensembles and the belonging field ideas and ideas are at rest, i.e., are endowed with the same type of motion. This means that they all obey the same logic of development and are subject to the same standards of measurement in the comparative procedures and ensuing judgements on congruence-coherence. Thus, every other idea, field idea, and manifold set of relationships between ideas not belonging to the principle idea but subject to its field influence when introduced into the same thought is like a body brought into the gravitational field of a cluster; it experiences a "deformation" and a change of motion until eventually brought to rest with the cluster's motion. By doing so, the previously "foreign" idea, and the correlated behaviors to which it may give rise are in turn experienced as deformed by an observer at rest in the "foreign" reference framework. The apparent illogicality of the relation A-C in the example above therefore rests on the fact that, when we let one of the many principle ideas guide our thoughts, we find ourselves in a reference framework having its own dominating logic; this means that in it the relationships among the elements of reasoning are measured by different metrics.

From the preceding arguments we may draw two orders of conclusions. On the one hand, in each of those principles that we call broadly philosophical there is an inherent logicality; each consists of a coherent system of congruent relationships. Furthermore, the intrinsically coherent system of eidetic relationships, once recognized as being independently self-supporting, has the same status as any other physical system of relationships —such as the cosmic one from which the analogy was taken. In fact, the thus recognized principle idea allows predictions about brain function, particularly about eidetic and behavioral aspects, that are as reliable as those about the motion of physical bodies and their electromagnetic phenomena in a gravitational field. On the other hand, we cannot deny that on purely matter-of-fact grounds no one would be able to detect a logical connection whatever between phenomena so far apart, for instance, as the deviation of star light by the gravitational field of the sun and the aforementioned refusal to drink despite thirst. However, when we start brooding over the possible framework of principles within which the drink refusal may be logically explained (i.e., when theorizing thought leads deeper into the essence of the phenomena), we stumble across very close analogies between the laws of the immaterial and those of the physical worlds. While perhaps not providing a unique explanation of the biological phenomenon, these may help to frame it by recognizing the presence of the same "structure" in the different situations, thus unveiling the same intrinsic significance of different events. Obviously, the similarity between biological and physical laws need not be restricted to similar formulations (although we could actually state them in the same way) but in the fact that, although each one is valid in its own region, there is a common essence, so that they may be unified at this level by the intrinsic similarity in logical structure (i.e., of the principles; see *PTN*).

We still need to connect these concerns with the preceding discussion about theorizing and analyzing motion. We have demonstrated that theorizing motion presents clear-cut correspondences between mind's and matter's operating principles and representations of the events and suggests ways for ordering these in a system, i.e., a theory can be outlined. A theory is a timeless, unitary framework made up of constituting principles, obviously all simultaneously present, comprising all the elements and the whole system of relationships linking them. This framework, once it has become the object of thought, i.e., embodied in brain matter, is not simply a system of coordinates, a new space within which the brain plays its role. It is the brain itself which comes to constitute a new reference space, comprising all the physical and nonphysical elements, the abstract objects and their relationships, i.e., the eidetic objects and their embodiment as physical forces; likewise, the physical space that we inhabit is made of matter, time, and energy, i.e., matter and its relationships. In this concept of space, the abstract and the physical one and their combination, the principles and all the abstract objects and relationships encompassed by the principles are simultaneously present. A theory is not a box with a content; it is a structure that cannot be split into container and content.

It must be stress that a theory is a full body of knowledge, complete in itself -constituting both processing capability and lines of development. The latter may not be fully clear at first sight, but they are nonetheless part of the theory and must only be recognized at a further enquiry. A broadly philosophical theory with these essential characteristics does not differ from other theories of a mathematical or physical nature. The latter also constitute a full body of knowledge, albeit with still incompletely explored lines of development; however, provided the theory is consistent in its fundamental principles, the scientist is guided by these principles in the recognition of still unknown regions. Within the theory space, then, after its main body has been outlined, we may proceed with the usual analytical procedures typical of the space-time that we live in. We are bound to do so, not because the abstract objects of mind's space are organized in terms of before and after —they only recognize a timeless, logical antecedence and consequence— but because our mental processes using language (itself physically embodied in neuronal circuits) are so organized that we can explore only according to temporal sequences, revealing the latter's particular "consequentially" logical or cause-and-effect, relationships. This is the unique and —alas— imperfect, way at our disposal for representing in our space-time what exists only as a system of simultaneous relationships between objects in the organized mind's abstract phase space constituting a theory. (Below we consider the way in which will can interfere with eidetic activity.)

5.4 Will and Eidetic Operation

In this section we refer specifically to the dynamic of reasoning as it is normally subjected to control or guiding by the voluntary activity of mind. It is a common experience to initiate and then change the course of a reasoning through a voluntary intervention, i.e., interfering with the eidetic processes by a decision independent of them (at least apparently). We consider now the modality of this interference and its limits. The problems concerning will are considered in the "Appendix" of *PTN* in an essentially philosophical and ethical context. Here, the matter is considered more in relation to brain function but without losing sight of phenomenological implications (to be investigated in Chap. 10). Will may manifest itself in a number of different ways, for instance, by starting or interrupting a reasoning, a motor act, sometimes also the secretion of glands and the movements of inner organs (bowels, bladder, etc.). However, to avoid dangerous confusions, we to separate carefully the processes leading to decision which constitute will (what we may call the "internal" will) from externalizations of will in the acts listed above. We shall be concerned essentially with the former.

We should first answer two questions: what is will, and how it is generated in our brain. The first of these is more metaphysical than neurophysiological, but we shall begin with an empirical, observational view: will is a decision that may or not be externalized in behavior. As a decision, it is the conclusion of a line of reasoning, a rational process, and is therefore a rational construction itself. Although one may not feel it to be an exernalized intention, i.e., itself in a pure eidetic condition, in its proper, commonly held sense will is an *act* of decision, the conclusion of mind procedures that realize the purely eidetic intention and therefore observable only through its effects, commonly the acts of the induced behavior. In this sense, will —as well as many other constructions of eidetic activity— resembles a physical force function; it too may remain in a potential state (and as such be measurable only if allowed to produce effects, that is, only if tested in some way). Of course, as may be expected, its nature is not physical energy itself; rather, it can only be related to the eidetic energy, to the gradient, or difference, connected to different possible choices. We may view this as an operational definition of will, leaving undiscussed the question as to the cause and the how of that eidetic gradient. We have considered gradients in field ideas above, and we may generalize the concept to all eidetic processes. This is a wide-ranging concept; as a force, will can be prevented from exerting its effects only by dissipation of energy in overcoming resistance or by another, counter-balancing force. We must therefore consider these factors whenever concerned with questions of pending decisions (see below). However, with this nothing new is introduced into our picture of the eidetic function —in particular, no dualism of will and counter-will— because we are still in the realm of rationality, which weighs arguments for and against within a single, unitary function. Therefore, expressions such as "dissipation of energy" and "counter-balancing of forces" are still to be considered as mere descriptive, physical analogies, without any implications (for the time being) as to the real mental processes.

We now examine the problem of modality, that is, the way in which will can intervene to modify or to interfere with the eidetic processes. This is of no minor importance in view of the discussion above concerning the heuristic procedures of analytical and theoretical thought and that of cause-and-effect relationships in eidetic operations. A line of reasoning follows a logical development that depends less on the specific object than on its mode (analysis, theory) and its nature (logicomathematical, complex). It has therefore been stressed that the conclusions of any reasoning are implicit to a greater or lesser degree in the premises of the reasoning itself, even if not clearly intelligible before as to their logical development. This is also the case for each intermediate step because each step is represented by a proposition ending at a point at which a decision must be taken; this means that each proposition in some way calls for a conclusion and thus a choice among alternative lines of development. In only very particular instances do steps not require such a choice, the single appropriate conclusion being clear from the beginning. Generally, mind feels its way forward in determining the best choice. Obviously, wrong choices are sometimes made because of mind's inability to detect the appropriate alternative.

Each choice is thus the result of a reasoning and is therefore a rational act; it also requires a voluntary act. Will generally complies with the rational judgement and the logic of reasoning. However, mind's voluntary intervention can interrupt the chain of logical choices at any point to favor another line of reasoning. Moreover, the decision to interrupt a line of thought, even if apparently unreasonable, obeys another logic and is the expression of the will claiming its own freedom (see Sect. 10.3.1). Although as a purposeful act will can divert the reasoning, in no way can it "alter" it, regardless of whether it is of the purely logicomathematical or the complex type. That is, will cannot change the logic of the eidetic process; it can only influence the line of reasoning in one direction or the other. As long as mind believes to be following logic correctly throughout the chain of choices, will complies with a rational principle at each step, even if mind is actually wrong (and, thus, the results at the end seen as nonrational).

Mind's ability to follow the steps of reasoning can become impaired only by decay of the rational capability itself since the steps are logically ordered not only according to type and mode of reasoning but also by the nature of the object of thought. Mind's task is generally to uncover and recognize the logical order at each step, and this depends on mind's intrinsic capacities. These of course depend on the individual's own characteristics, such as training in abstraction processes, previously acquired knowledge, richness in internal language (see Chap. 10), interest in the object of thought, etc.; however, none of these properties is dependent on will. Therefore, once mind has recognized the logicality of the subsequent steps, will is left merely with the choice of either accepting it (i.e., complying with the logic of thought) or directing reasoning toward other lines; again, this is a choice based on rationality (see Sect. 10.3.1). Thus, will is always associated with rationality and therefore with the eidetic process. We could phrase this more accurately by saying that in view of its continuous involvement in each step of reasoning will is closely bound to each eidetic process, not as a function depending on eidetic operation or as a product of it but as an essential part of it.

From both the psychological and the neurophysiological standpoints we may see will as determining every nonautomatic behavior, including each decision step during the reasoning procedures.

Our position maintaining that will cannot alter the logic of reasoning, is not one that can be disproven. Nonetheless, there are different ways in which will can interfere with reasoning other than directly determining decision steps. We have paid little attention to complex reasoning, that involving mathematical-scientific elements together with broadly philosophical ones (ethical, religious, strictly philosophical, social, political, etc.). Everyday life seldom involves purely logicomathematical formulations, and mind is almost always concerned with the broadly philosophical elements that guide thought and behavior; therefore, both reasoning and the resulting behavior are continually submitted to explicit judgements on logicality. Neither as elements entering judgement nor as factors causing a change of reference framework can these broadly philosophical elements disturb logicomathematical reasoning or alter the predictability of mathematical steps; the solutions of purely mathematical problems do not change regardless of the framework in which they are considered. However, in all other cases the introduction of elements transforming the reasoning into a complex one is likely to produce differences in the processes leading to decisions. As we have already stressed, the broadly philosophical elements may induce a change of framework, and in different reference frameworks there exist different (local) geometries and therefore different standards of measurement. Since the decisions depend on comparisons between measuremeents performed with the local standards, it is possible for the decisions to be different in different frameworks if there has not been a uniform transformation of the space and time parameters. In this case, every ensemble of parameters that is not strictly logicomathematical could exhibit different local standards of measurement, and therefore reasoning could reach conclusions —and therefore resulting decisions— apparently in contrast to those logically expected in the absence of transformation.

This situation of contrasting logic (and of possible conflicts between decisions) can be avoided only when the broadly philosophical elements enter the new reference framework explicitly as new parameters and therefore also as new elements of judgement. As a consequence, that which may previously have appeared as the most logical decision can now be voluntarily blocked in favor of another different, and sometimes contrasting, operative decision —which, however, in the new framework appears as logical as did formerly the blocked one (in the previous framework). Only the presence of broadly philosophical elements of reasoning (and therefore of judgement) could induce the change of framework justifying on perfectly rational grounds the refusal to drink by a thirsty man (see Sect. 5.2). This voluntary refusal is not an alteration of the logic of thought and therefore does not indicate a power that will actually does not possess. Rather, it is the result of a comparative measurement between all the relevant parameters performed in a space in which the local metrics has changed. The relative scarcity of the available water for all thirsty men is endowed with much greater judgemental weight than the thirst itself; will decides accordingly and directs behavior toward another solution.

On the other hand, within each framework there also are sets of elements' the weights of which covary, and decisions dependent on these sets cannot be changed in any way —this is the case with purely logicomathematical elements of thought. Here too will is unable to change the path of reasoning, although it can redirect thought by selecting among various broadly philosophical principles, interests, or lines of thought. The logicality of subsequent steps is ensured only if the suitable framework results from transformation according to definite lines of thought, principles, or needs. This corresponds to a voluntary change of reference framework and thereby to the resulting selection of decisions which could not be logically taken in the former framework.

In the "Appendix " to *PTN*, I discuss the freedom of will in selecting the guiding principles of our actions, and here we have reached the same conclusion by considering the gross phenomenology of the relationships between will and thought at the origin of behavior. In that "Appendix" our actions are seen as always being determined, in the sense that we always obey some principle and are therefore free only to choose the principles that we wish to obey. Thought and behavior always follow a general scheme, a preordered pattern which imposes constraints depending upon the specific nature of the particular principle. The various different sets of broadly philosophical, scientific, and mathematical logical elements are so entwined that it is difficult to identify preordered patterns clearly related to only one or the other of the sets of elements. Nevertheless, operative decisions are at least to some extent predictable because within a given framework the basic logicality of the line of thought is itself preordered. Being able to recognize the logicality of a line of thought requires that one is able to direct his own thought toward that framework, i.e., to penetrate and explore that space. More properly speaking, mind must voluntarily accept the corresponding local metrics and the local standards of measurement, which is another way of maintaining that mind must operate a transformation of reference framework. In other words, mind appropriates the logic of each reference framework, but this requires knowledge from vastly different regions of thought.

We must also consider the action of will in neurophysiological terms. The active intervention of will has the appearance of inhibiting one line of thought to facilitate another. However, from the above discussion it is clear that one cannot speak simply of inhibition or facilitation, particularly not in neurophysiological terms. The change of reference framework, though conceptually simple, is an operationally complex procedure, involving (according to the theory of *PTN*) a "neurophysiological" metric tensor, as yet not identifiable with the simple operations monitored through microelectrodes in individual cells, i.e., the usual play of synaptic inhibition and excitation. The voluntary choice of strictly logical and scientific principles or of any of the broadly philosophical ones automatically brings the transformation of the reference system corresponding to the selected principles. It is very probable that the diffuse projecting subsystems of the brainstem and thalamus are involved in the actual operations of transformation, but the organization and operative modalities of the procedure at the level of the different sets of neurons are completely unknown and constitute by themselves a challenging task for the investigator (of course, at the level of cell activity one may observe only the fundamental operations of excitation and inhibition).

5.5 Will and Emotions

We begin this section with a summary of the role of the broadly philosophical elements in eidetic processes. These can be considered in tree ways: a) as weights, in the sense that their mere presence adds algebraically to other elements and alters their significance; b) as rational elements of judgement, entering the comparisons and thus the decisions during the eidetic processes; and c) as constituting frameworks of reference (when considered in their various ensembles) possessing their own geometries and thus local standards of measurement and therefore possibly inducing different logics of judgement. A common example may clarify this point. The belief in the immortality of the soul and rewards in an afterlife can attach a weight to an otherwise unbearable evil that makes it bearable. This belief, in the context of rational elements of judgement and a broad horizon of knowledge may induce the voluntary, logical choice to suffer an evil in this life as a (quasi-necessary) condition for enjoying a superior reward in the next. Such concepts, as part of a principle idea, entail a transformation of mind's reference framework suitable for taking decisions and for organizing future behavior in accordance with them also in all other situations of life.

Another factor closely related to the problem of will and its freedom is emotionality. We set aside, first of all, "feelings," the personal way of living the emotions. Emotionality is expressed through a number of physiological features in the autonomous system and its dependent organs and is observable in a complex of behavioral manifestations. These are controlled by an ensemble of neural structures comprising basal forebrain, hypothalamus, parts of the thalamus and of cerebral cortex, rhinencephalon, and of course structures of the autonomic system. In this large complex, closely connected with the rest of the nervous system, emotional feelings are integrated with the autonomic and somatic reactions characterizing each emotional state. Of the greatest importance here is the way in which one lives the emotion, because the one that lives the emotion is the psyche, an entity that is even less clearly defined than the eidetic structure itself. The content of emotionality is extremely varied, and its relationships with the eidetic processes are correspondingly complex. The conditions of life today have certainly added more weight to feelings such as anxiety, psychic excitement and infirmity, and to phobias, sex pulsions, and greed of various types, all of which are added to those of the common emotional categories of love, hate, jealousy, rage, joy, fear, etc.

What concerns us here are the possible ways in which the emotion-evoking elements may interfere with eidetic processes and will. The emotions are commonly thought of as having the opposite character to rationality —nonrational or even irrational. Emotions are physiological manifestations of psychic life; some of these are common with animals, although there is no clear-cut demonstration that they have psychic features in animals comparable to those in humans. The human brain is extremely more complex in structural organization due to fusion of philogenetically older and newer structures to form functionally more powerful and sophisticated complexes; in *PTN* I maintain that whatever functions may appear to be common between animals and man are performed by man in a more complete and an essentially different way. The emotions are essential and indispensable physiological factors of human life; they are integrated and regulated within an ensemble of brain subsystems, giving rise to organized behaviors with largely life- or species-preserving importance (a primordial aspect in animals, but in man overshadowed by his characteristic psychic features). Even disregarding clearly pathological conditions, the emotions exhibit obviously varying degrees of intensity without losing any of their strictly physiological character.

By virtue of being physiological in nature, emotions and feelings belong to a logical and thus rational overall network of brain activities and are therefore themselves logical and rational. However, the important point is that, as emotions, they do not constitute reasoning in themselves, nor do they depend upon reasoning. They are not the result of thought and in no way form part of an eidetic operation. In fact, in the case of most of them —for instance, phobias, fear, rage, hate, love, jealousy, anxiety— a rationally conducted physiological analysis of the related feelings can divest them of their overpowering and urgent character. This leaves behavior more freedom to follow the rational decisions of thought, based on the appraisal of the real situation that has given rise to the emotion. The most emotion-loaded elements are those centered on the self and the own good in a broad sense; they are therefore philogenetically rooted in the logic of the human's constitution as a living being. The primitive nature of most emotions excludes —we might say, by definition— that they may have any connection with eidetic processes, although they belong to the rational network of brain organization, and their associated feelings may be submitted to rational analysis, i.e., are rationally intelligible.

If the emotions are not the result of reasoning, they cannot be created or evoked within the eidetic structures and can therefore interfere with eidetic activity in only two of the modes described above for the broadly philosophical elements, that is: a) as weights adding algebraically to the weight of the other —rational— elements and b) as factors causing transformation of the reference framework in which reasoning and behavior are developed and organized. On the other hand, since the emotions cannot represent rational elements of judgement during reasoning, they cannot direct or guide reasoning or contribute together with the rational elements to the rational decisions —the sole basis of thought and part of its essence. "Weight" is perhaps the more appropriate term for indicating the overpowering and urgent character of the emotions. These, on the one hand, seem massively to impel our behavior toward a goal without the apparent intervention of reasoning; on the other, they are seemingly able to constrain thought to circumscribed interests. We may envisage this constraining power as linked to the strong attraction exerted upon mind by a "heavy" single object, so that the eidetic processes become concentrated on that object. The convergence of the two factors, the automatic components of behavior (somatic and autonomic, typical of the emotional expressions) and the restricted horizon of thought, indicates that the emotions, though noneidetic in nature, may nonetheless predominate and severely interfere with a smooth progression in eidetic operations. Let us now consider more closely the significance of this weight.

We have said that the weight of the emotional factors substitutes for rational elements in determining the behavior but not the eidetic processes as such, thus not really interfering with these processes in the sense of altering the rational procedures of comparison and decision. The sole direct interference with eidetic processes lies in the restriction of mind's attention, either to extremely narrow areas or even to the specific object of the emotion. An obvious and unavoidable consequence in the latter case is that subsequent behavior more easily follows automatic patterns, specifically those entailing species- and life-preserving features, such as flight before an impending danger or the aggression of enemy or sexual partner. The automatic patterns can therefore be seen as built into the structure of the brain circuits (as the hardware of a computer) and representing the philogenetically older, fixed organization in terms of which animals and humans behave to some extent similarly. Nevertheless, we should not think an emotion-driven man to resemble an animal or a decerebrated man in being unable to use his rational powers. Since in humans the philogenetically older structures have fused with newer ones, we can never expect the automatic reactions of humans closely to mirror those of animals. The anatomical and functional links between the structures controlling or eliciting the automatisms and those supporting eidetic processes are also such that, even in primitive reactions, humans are distinguished from animals by the characteristics of reason.

When emotionality succeeds in restricting mind's attention to a unique object, for instance, in the case of hatred, the rational processes are not short-circuited, and the behavior still results from rational elements; it cannot generally be fully automatized although it is heavily conditioned by the content of that emotion, which may be directed only to the satisfaction of the drive associated with that emotion. Thus, far from consisting only of automatisms, the behavior develops essentially in the rational search for the way to satisfy the emotional drive and in the eidetic elaboration of a justification for the desired behavior, i.e., for voluntarily pursuing a goal that reason still does not recognize as properly of human standards or sees as altogether unethical or unlawful. Therefore, the interference of emotionality with eidetic activity consists of depriving the latter of the larger horizon of objects necessary for developing thought in a way suitable for reaching an unconditioned judgement, while leaving unaltered the ability to use the basic operations of comparison involved in reasoning and those not aimed at the satisfaction of that drive.

In these quasiphysiological conditions of mind —i.e., preserved, not short-circuited rational powers— will can always be considered as free and responsible. The question, however, is actually not so clear-cut and deserves more accurate analysis. Let us begin with the second type of interference of emotionality in eidetic activity, i.e., transformation of the reference framework caused by emotional factors. It is obvious that this is entirely congruent with the modality of interference between emotion and eidetic processes as examined so far. The basic character of this relationship is the narrowing of the "eidetic spectrum," with the consequence of centering the reasoning predominantly upon the object of emotion (no evidence has so far been presented of a narrowing of the whole eidetic space). A transformation of the reference framework can be presumed not to be uniform, so that the new reference is represented by a strongly deformed space. One cannot say that the space coordinates concerned with the eidetic processes have disappeared because as we have noted both emotional and rational elements depend on strictly interconnected nervous structures and are always present (indeed, one cannot imagine how the rational elements constituting the basis of eidetic activity could possibly disappear from the cerebral space). These may strongly compete with each other, however, for the control of behavior, since they are loaded with comparable weights. Substantial differences in the weights of the coordinates (if possible) would result in virtually excluding a number of eidetic parameters, entailing in turn a reduced number of choices in reasoning and thus eventually depriving the judgement of the necessary elements on which it must be based. In this case, the freedom and responsibility of the individual could be questioned. Leaving aside the philosophical, ethical, and juridical aspects of the problem, maintaining a fully rational and independent judgement can be ensured only by thoroughly training the eidetic function, as it has been hinted in earlier sections.

The operational side of the hypothesis requires the neurophysiological (and psychological) mechanisms of attention to be diverted away from the eidetic processes —which can lead only to rational decisions— toward those processes that lead instead to automatic reactions. As discussed in Sect. 2.3, interest-driven eidetic attention is seen as the factor determining the reference space transformations. Analogous attention-related mechanisms may be responsible for the relevance assumed by the emotional elements in the corresponding transformation of the space. In *PTN*, the strong intervention of emotionality is viewed as a likely basis for space transformation, explaining the time in emergency situations, interpreted in the relativistic sense of a slowing of the personal time and thus of the time base of the neural processes. Also in animal behavior it has long been demonstrated that particular attention-attracting stimuli, even if restricted to the vegetative sphere, can produce a generalized modification of the balance of weights associated with other neurophysiological elements, so that the whole system of the quantitative relationships among them is changed (which amounts to a change of reference system), with resulting unexpected, though in themselves logical, modifications of behavior. The example above (Sect. 5.2) of the behavior-conditioning broadly philosophical principles may also be seen in the present light as that the religious, ethical, social, and political elements in these principles may also contain sufficiently strong emotional drives.

On the other hand, the narrowing of the eidetic field may be more apparent than real, even if observations seem to suggest it, when one considers the peculiar character of the subspace spanned by the emotionality variables (as elements of the subspace). It is quite reasonable to presume, as we have just done, the emotionality subspace to posses an enhanced attractional (gravitational-like) power; in terms of our theory, this means that a strong force is exerted by the emotionality subspace on other subspaces connected to it. Since there is hardly a subspace in the brain that is not connected with all the others, we must expect that the deformation of space geometry, induced by the emotionality subspace "strains" the coordinates of the other subspaces as well, introducing changes in the course of the neural functions represented there. However, since reasoning, although centered on a single object, in itself cannot be altered (as we have seen), the change cannot consist in an alteration of the mechanisms of logical comparison and decision of reasoning. This suggests that the coordinate variables representing the eidetic function may also be strained by the same amount, therefore maintaning unchanged their reciprocal relationships. As a result, a change would appear only in the "operative decisions," those which decide the final, resulting behavior. At this level, a true rivalry may appear between emotionality, on the one hand, and rationality and broadly philosophical principles, on the other, giving access to a number of possibly different and contrasting manifestations.

We should consider that the same emotion-evoking situation does not alone determine a specific behavior. For instance, the choice between aggression and flight does not depend solely on the situation itself but also on the existing brain state and preceding history; therefore, it may not be completely determined —and thus predictable— even in animals (although in the latter rational choices are barred for want of the necessary supporting structures and related organization). In humans, therefore, it is the more likely that rationality is still the real deciding factor by favoring either of the possible behaviors on the basis of comparison and of predicting the consequences of the selected action. That the selected behavior may nonetheless prove inadequate for the situation or simply not to represent the best choice from the rational point of view is probably more the consequence of a necessarily hasty decision —as required by dangerous situations— preventing reason from suitably taking into account all the arguments for and against. The correct "use" of free-will is preserved in any case.

Let us now consider emotionality and (eidetic) rationality as plain forces. The peculiarity of the emotion-driven behavior is largely due to the fact that emotions are always accompanied and partly determined by neurally and chemically evoked secretions (from pituitary, adrenals, and other glands) as well as by release of mediators controlling the bulk of catecholaminic pathways in the nervous system. Since each type of emotion can probably be characterized by its own "chemical spectrum" of secretions and mediator molecules (although no conclusive, direct evidence is available), the rationally driven decisions must face the whole neural and chemical organization of the emotionally determined ones. Therefore, the operative decisions would only be taken on the basis of the ratio between the driving forces inherent in emotionality and those in rationality. In analogy to neural motion, based upon analogous physical laws, the deciding factor is the ratio between the rational and emotional momenta of motion. From this we may conclude that the effective operative decisions in humans depend on the momentum components represented by factors such as eidetic training, culture, and knowledge with its horizon, all of which contribute to the momentum of purely rational motion and eventually succeed in effectively opposing the motion of emotionally driven automatisms. Therefore, in no case is the intrinsic power of rationality impaired since the capability for conducting and developing the eidetic processes would be out of reach of the factors belonging to emotionality. On the other hand, it should not to be wondered if, even in subjects thoroughly trained in rational thought, a sudden situation with a strong emotional content gives rise to completely automatized reactions. The unusual character of the situation, together with the necessity of quick decision, could find an otherwise cultivated and rationally trained brain with an insufficient knowledge of emotion-dominated situations and therefore unable to rationally master them at their first occurrence for want of previous experience and training (not simply at the real but also at the imaginative level). The freedom of will is always ensured whenever the eidetic function and its rational procedures have access to the decisional mechanisms (the most general case in everyday life). Only in extreme situations (for instance, life-threatening events, such as in street traffic, in war, or in other catastrophic conditions) only the will-determined action —and not will's freedom of decision— is barred, since action and not pondered decision is produced by the emotional "force" liberated by the life-preserving necessities, which may leave little or no time for the development of the slower rational weighing. However, these instances also still belong to the range of physiological behavior, for they entail reactions, albeit automatized, that are directed to the rationally identifiable supreme "good" —life— much like will, as we shall see below, with its well-aimed decisions (see Sects. 10.3, 10.3.1; and *PTN*, "Appendix").

5.6 Will and Compulsions

In catastrophic events man's operative decisions and thus his behavior may cease to be predictable and ostensibly determined by free will as the situation may be beyond his rational capacity to cope quickly. However, there are other instances (apparently outside the range of organic brain pathology and giving rise to the mere automatism of either patterned or random muscle contractions) of seemingly coercive behavior and presupposing prior coercive operative decisions. Coerced or compulsive in this context may be defined as those acts which seem to be organized outside the emotional drive and against the subject's will, in other words, behavior not resulting from an actual choice and therefore not necessarily directed to the attainment of a desired good or to the satisfaction of a conscious drive. This ought to indicate that in these individuals no truly rational choice or no choice at all is possible; however, in all other aspects of everyday life the same individuals would seem able to conduct reasoning to an end and to make the correspondingly correct behavioral choices.

Absence of choice in these compulsive behaviors may occur for two reasons: a) the overwhelming weight of some factors that may prevent other elements of judgement from exerting their own weight in the eidetic process (the case of strongly different momenta of motion, see the preceding section); b) a substantial deficiency in the logical thought procedures. The latter seems to be excluded as a generalized cause of compulsion since outside the specific compulsive behavior most other acts may be perfectly justified on rational grounds. Moreover, such a profound disturbance of logical processes would easily be detected by relatively simple tests (this may easily occur in senile decay, although compulsions do not seem to appear more frequently in pathological aging). Until severe deficiencies in the thought processes are uncovered, the presence of overwhelming weights barring rational choices and therefore imposing irrational behaviors ought to be seriously considered. The actual mechanisms of such nonrationality might consist of some mechanism barring the transformation of brain state into the suitable reference framework. For instance, when ethical principles are involved, the unsuccessful transformation of the existing reference framework into the ethical one may result in shameful activities (as is frequently the case of compulsive acts). However, since we previously concluded that the choice of reference framework is the true voluntary act, and within every framework the development of thought is guided by the logic of that framework, the compulsions seem to reveal defects of will —or,better, defects of free will. However, the fact that in the extremely wide range of correctly will-guided behaviors, only one act or one type of acts is not submitted to rational judgement should make us suspect that we are not confronted with truly faulty mechanisms of eidetic activity or of framework transformation, nor of the ultimate, rational, operative will-dependent decisions. Rather, the origin perhaps must be sought in the more profound psychic aspects of brain life, which may secondarily interfere with the former. It may even be that the compulsive act actually *is* a voluntary one, originating at the basis of will and therefore rational and externalized through the normal voluntary channels; however, this is an act that the subject refuses to recognize as voluntary —being fully aware of its futile and even shameful content —and therefore simply refuses to take notice of (see Sect. 10.3).

6 Resumé and Comments

Before closing Part III, we should summarize the main lines that have been developed thus far, as Parts I-III constitute the basis for Parts IV and V, which are concerned with the essence of the eidetic function, ideas, and their mysterious ways of existing in the brain. We have maintained a separation between the heuristic procedures of analysis and those of theory. A theory is a structure created by mind but aspiring to represent the whole set of relationships among ensembles of element-ideas belonging to a same phase space, or field, of science, whether in physical or in immaterial reality. In this context, the term "created" may not always appear the most appropriate, as the complex system of relationships in Nature —including all its objects and their reciprocal relationships—has self-sufficient existence independent of mind. Nevertheless, mind is part of man, and man is part of nature. The self-sufficiency of nature extends to mind, which creates abstract structures to order the system of relationships among the objects of nature. We have noted that the abstract structures created by mind are the only ones that mind can understand. Mind gives form to these relationships by using symbolic language which alone brings order to the elements of the phase space and makes it understandable to others. However, since all the relationships among element-ideas are not easily and immediately understandable, mind interprets that order —this means the creation, not merely recognition, of the logic of matter-of-fact situations. Drafting a theory thus means composing a unitary picture of the hidden logic binding the abstract structures that are behind the observables of nature. To the extent to which the theory coincides with the "real" logic of nature, it exhibits an independence from the mind that created it.

In studying the phenomena in any field of science, the researcher seeks not only to analyze partial and particular events but also to abstract a logical, general order from these events. We have argued above that there is no logical procedure leading "backwards" from facts to theory. Reasoning that originates in factual reality consists merely of low-grade induction, and the heuristic principle involved is the tacit and arbitrary assumption that a law may be based upon the ascertained frequency of facts. However, we concluded above that inductive reasoning is twisted reasoning. Certainly, the existence of some type of order may be demonstrated by the fact that the results of the experiments are repeatable, at least under controlled conditions —actually we should say that the repeatability rests on the existence of some kind of order. The nature of the order, that is, what general principle is involved, cannot be inferred from any type of experiment, since the results of all possible experiments, considered individually, are related to only partial aspects of the general framework (the theory). In this way, the low-grade inductive procedure presumes that the underlying mechanism of an event, once the results appear sufficiently stable in a statistical sense, is all there is to know and is itself the law, therefore a theory (although it can represent only a dependent product, practically resulting froma law, and is subject to explanation by a theory). As was noted above, these so-called laws cannot represent anything more than a reformulation of the results of experiments and cannot claim to represent general principles, since their own domain and their efficacy remain confined to that type of experiment. Furthermore, since these laws are essentially tautologies, experiments based on them can produce only self-confirmation. There is also another, more hidden point. Since the conclusions drawn from this type of experimental principle are based on some frequency of results, they contain an undefined degree of uncertainty: first, one arbitrarily introduces statistical reasoning where there is generally no need for this at all, and then, because the level of statistical confidence is also arbitrary, there is no true and unique guiding criterion applicable for every case.

There is also a psychological danger here. With his attention concentrated on the results of experiments, the investigator loses sight of the larger ensemble of interconnected problems within the reference framework as represented by the principles of that field of science. This leads him to consider the events and facts as he observes them to be the essence of the things and to constitute the only real problem for mind. This illusion is particularly dangerous. Both in the speculative and in the practical aspects of research the most fruitful and effective contribution of the investigator lies in his insight into the framework of principles underlying the physical and the biological mechanisms; this is at a point far removed from the actual experiment but is nevertheless the correct way in which to conduct research. Experimental research can concern itself only with investigating the specific mechanisms involved, whereas the general framework makes no reference to experiments but is an abstract structure that predicts the outcome of a great many different types of experiments, and depends on no one of these in particular —it is created by a logic independent of the sequential logic of events in the space-time of nature and thus enjoys an autonomous existence. Far from being an ensemble of partial laws, one for each particular fact, the general law —the theory— comprises a unity that does not require a great number of experiments to test its truth. A characteristic of general frameworks is that they help us to see in advance. Rather than looking backward to experiments, a theory looks forward; it is the product of the freedom of mind from strict time consequentiality, from the law of before-and-after, of the cause and the necessary result (see Chap. 3).

To construct a theory one starts from some analogy, either borrowing a fully developed theory from other field and adapting it developing a new one. Probabilistic considerations may enter into the original selection of the starting point, particularly when the analogy is sought in the same or in a very nearby field. However, by its very nature, the truly analogical procedure for building a theory is the more general —and thus effective— the less it is mingled with probabilistic calculations, that is, the less it becomes entangled in the results of experiments or in observed data. This means that analogical reasoning aimed at development of a theory does not take the backward path from effect to cause of a specific event because it would hardly find the same specific mechanism for different events —in other words, the phenomena in different fields would not exhibit a similarity of mechanism (which is implicit in their being the outcome of the action of factors in different fields). This entails in turn that analogical reasoning is effective only at an abstract level and must seek the hidden, intrinsic, common meaning of the events of the two fields. Therefore, there are varying degrees of analogies. The lowest is built almost exclusively upon probabilistic considerations deriving from previously collected specific evidence in the same field. Higher grade analogies are those in which no specific experiences are involved but the more abstract meanings of events in different fields of science. From these meanings stem the more general laws, which may be transposed as general principles from one to the other field and having exactly the same bearing in both.

In analogical reasoning mind must rely on its own abstracting powers, not called for at all in the case of the lowest grade analogies but certainly important in the task of recognizing higher grade analogies. At the basis of this is the judgement on congruence-coherence, which decides whether the analogy really exists —an analogy test applied not to single ideas but to sets of them (better, to whole reasonings and situations) considered in their abstract meaning. As a matter of fact, following this procedure, congruence-coherence must be found between the postulates and the predicted phenomena of the two theories. Put in these terms, there is no doubt that analogical reasoning may lead to the formulation of a theory; however, if one had to rely exclusively on this, a "ready to use" theory should always be available. This could refer us in turn to another theory. To escape from endless recourse to analogies we must be able to draft theories through other procedures as well; indeed, reasoning has other, as yet unexplored facets.

Etymologically, inductive reasoning is a procedure that takes observed facts as instances of a general law (which we reject on principle) or takes specific situations or theories in nearby fields as examples (i.e., analogies). In a case in which the mind participates more actively it uses knowledge not from only one other field (i.e., a straight analogy) but that from many different fields. Mind can sort these elements out, and make of them a unified, integrated set of principles to be tested in experiments. This procedure rests on the ability to recognize the congruence-coherence among abstracted elements from different regions and to assemble them into a new general framework. This mode of reasoning differs substantially from borrowing a ready-made, fully outlined theory. It requires of the mind that it be able to create. However, for procedures seeking the right path through a maze of elements from different fields, particularly when lacking a preexisting model (or pattern) the term intuition is more appropriate. This involves the ability to understand systems of relationships not previously thought of and without examining the details of a problem; it is the immediate grasping of the essentials. Thus, it is a mode of thinking in which self-conscious, stepwise exploration plays little role —in contrast to "flashing" thought.

Intuition involves an eidetic process that is difficult to comprehend for it apparently takes shortcuts through our everyday logical procedures of reasoning. It is obvious that we cannot dispense with logic in any way; therefore, its explanation must be sought in far more subtle processes than those considered so far. These will become understandable only when the problems concerning knowledge (discussed above) and those concerning consciousness, conscience, and time (due to the shortcuts and flashing character of intuitive thought) are dealt with (in Part V). The essence of eidetic activity is such that it cannot be treated following a strictly logical order and a strictly defined temporal sequence. Objects of abstract nature —ideas, field ideas, concepts, principle ideas— comprise many elements but are indivisible unities; they include no antecedent and result, but everything is present in its completeness, in a direct and immediate way. The hypothesis advanced above considers mind's products in the form of unified, indivisible theories in which all the elements are indispensable. The hypothesis that I am now advancing is that the shortcut process of reasoning, intuition, recognizes the whole when only a part can be seen, therefore a whole that in some way preexists in some space with all its elements, and needs only to be sorted out from its myriad abstract structures. Below we examine how this hypothesis fits with the larger theoretical framework of the eidetic function.

Part IV

Ideas

7 Stored Ideas: An Approach to Memory

In Parts I-III we examined the questions concerning thought and its dynamics; here we turn our attention to the idea, the main object of the eidetic function. As an introduction to the procedures of thought, we looked in Chap. 1 at the idea as a unitary, "condensed" thought that can be reduced to a symbolic formula. Now, we consider the ontogenesis of the idea in an attempt to identify its crucial aspects. In doing so, we also confront the question of memory, which has been referred to up to here only tangentially.

Idea and memory are obviously closely linked. If everything is memorized in a stable way and is quickly retrievable for use in thought processes, then upon retrieval memory and idea are one and the same thing. However, should we see the idea, once formed, as having to be translated into another form to become stable and then transformed back again to be used? Is it not possible that idea and memory exist in the same form? Seen in this way, memory is a secondary matter in idea formation; perhaps memory is a problem only if one sees its formation as a primary act, directly from elements that are noneidetic in nature (for instance, impulse trains from sense organs before their interpretation). According to the latter view, the idea is generated only later during retrieval from a memory trace, receiving in this process the fleeting eidetic form believed to be necessary in almost all brain processes, including thought. This view, however, accounts only for those ideas deriving directly from nature (natural ideas) and implies a different procedure of idea formation for nonnatural ones (constructed ideas). The questions associated with memory are inexhaustible.

We advance arguments to support the view that a) physically stable memory is not a godsend in therms of thought dynamics since a two-way transformation would become necessary every time an idea is used in reasoning, and b) a stable memory as depository of all ideas would become the source of stereotyped elements of reasonings, hardly suited to meeting the need of innumerable nuances of elements entering in human thought as soon as abstraction processes begin. Once we accept a close logical kinship between ideas and memory, the latter can be used to detect the existence of the former when a direct approach is not possible or to detect differences in brain performance between humans and animals. We do not seek to solve the complex problems of human versus animal eidetic operations; however, a few more arguments are advanced to discriminate between truly eidetic memory and "noneidetic memory"

7.1 Memory as Storage: Is Useful?

Not only intuitive but also matter-of-fact considerations condition the mechanisms of thought —from the formation of ideas to the development of reasoning— to rely on a permanent pool of readily available elements for use in all eidetic procedures. Memory is a function (if this term is ever applicable in a unitary sense to the great number of different operations implied in it) which remains completely obscure despite the variety of hypotheses about the possible mechanisms underlying it. To begin with, let us understand memory in its common and colloquial sense of storage, possibly stable throughout life, i.e., persisting in the face of the many intervening factors continually threatening to destroy it.

The first question we ask is whether memory in the form of a storage would be of any advantage as far as the intellectual function is concerned. The question may sound futile or rhetorical, but it is intended here to be provocative since we are accostumed to thinking in terms of items believed to be permanently memorized, such as ideas, reasonings, sentences, events, images, sounds, mathematical formulas, and knowledge in general. The concept of recalling from a memory storage suggests the existence of engrams (in some still unknown place and in an even more unknown way). This common, graphic representation is not supported, however, by any sufficiently credible evidence. In addition, the foundation of eidetic function on the mere mechanistic depositing and retrieval of a system of stably stored memories would reduce reasoning to the reproduction of stereotypes variously mixed and held together by syntactic conjunctions. Ideas and sentences would be mechanically, quasiautomatically, and immutably repeated whenever the suitable triggering stimulus is present. (Examples of this are admittedly all too frequent, but these can be traced to a lamentable poverty of language, absence of true reasoning or at least of its spontaneity, or lack of adaptation of eidetic procedures to situations requiring the creation of new elements of thought.)

Since much experimental research on memory involves animal brain, a preliminary question should be whether there are different types of memory, some peculiar to man and others to animals. Man's memories are made up principally of symbolized eidetic items (we consider symbols and language in Chap. 10), thus an intellectual memory inasmuch as the items have been filtered through mind's language. However, man's memories also concern objects that are not strictly eidetic such as images, events, and past sensory experiences in general; these are filtered through mind and subjected to a more or less complete and rationalized elaboration well beyond the stage of simple sensory experience, so that their retrieval is never disjoined from their internal representation in terms of man's language. Therefore, they belong to the same type of memory as do ideas.

However, together with this memory another seems to exist, which is shared by animals and humans. Animals exhibit mainly automatic and very poorly critical, or altogether uncritical repetition of behaviors (see also the "Appendix" of *PTN*). Many of these are present from the beginning of the animal's life, i.e., they are thought to be genetically determined; others have been imprinted early or have been learned later through parental teaching, imitation, or the individual's experience in his environment. For the animals these memories subserve a vital function, representing indispensable, survival- and thus species-preserving factors. This type of memory, then, is typical of animals and survives in man (if it does) perhaps only as a remnant of the philogenetic development —though integrated with his characteristically human eidetic memory.

There is a danger, however, in emphasizing the survival factor which this primitive memory represents, as this can lead us astray as to the real role of memory in human life. To the extent to which there is still a survival function, this memory serves less to protect man's life as part of a general equilibrium within the natural world than, by protecting his life, to allow man to maintain sovereignty over nature. This may seem a teleological and philosophical interpretation, but it is far from being so. Man's behavior is never automatically or spontaneously aimed at maintaining natural equilibria. Through the rational properties of his brain, man subdues the world around him for his own needs and imposes on it an equilibrium that suits him; any equilibrium that exists is therefore not the natural one but one expressing man's domination over nature. In Man, the philogenetic remnants of automatic behaviors can be of survival value only in coping with extreme, emergency situations, not with the everyday problems that require him, instead, to be rationally controlled and carefully goal-directed. Therefore, in assuring man's long-term survival in a subdued nature, fixed patterns of memories are of no use beyond the emergency-defense ones, not only of the inherited type but also those that have been learned.

7.2 The Nature of Memory

The type of memory that can account for early-appearing automatisms of behavior in animals, as well as for some individual peculiarities in man is almost certainly genetic in nature. This should not be taken, however, as a genetic storage of engrams but as genetic determination of the basic organization of the brain structure, i.e., of cell connections. It is conceivable that in the brain a number of neuronal circuits may be genetically preprogrammed to express definite behaviors of higher survival value in response to particular stimulus situations, assuring to newborns and (if remaining) to adults the immediate emergency mechanisms of reaction. While it is evident that a few behaviors are genetically determined, since the organization of patterns of motor acts and gland secretions can easily be preprogrammed in the brain circuitry as stereotypes to be readily expressed upon presentation of the specific stimulus, we cannot assume that everything is determined by genetic mechanisms. We have no experimental evidence that patterns of visual, auditory, olfactory, or tactile nature can be genetically determined in such a way that behavior is dependent exclusively on them. Quite the contrary is the case. Imprinted phenomena, for instance, actually exclude a uniqueness of mechanisms, since they demonstrate that individual- and species-survival reactions are strictly dependent on stimuli received after birth. In these instances and perhaps also generally a genetic determination can still be presumed, however not as a specific stimulus-response pattern at the level of preprogrammed responding circuits but as a genetically determined readiness or capacity to acquire stable response patterns. Moreover, experimental psychology, by evaluating test results only on the basis of statistical probability, cannot demonstrate true invariance of these types of reactions in widely varying conditions —as whould be implied in the hypothess of exclusively genetically coded patterns of situations or of imprinting-generated, emergency reactions.

However, one might always suspect that similar types of chemical coding are at the basis of both genetic and brain memories. In nerve cells —as in every other cell— it is possible to construct an almost infinite number of different combinations of amino acid chains which may enter coded sequences for each of the innumerable items to be memorized; This would therefore make such a storing medium available. However, on the one hand, no evidence of actual coding and storing of memory items, besides genetic (and virus) material, has been provided (although it may be in the near future). Genetists, physicists, and neuroscientists, cannot as yet posit any reasonable hypothesis about the functional organization of the two-way chemicophysical and neurophysiological processes through which a memorizable item can first be translated from the original form of activity of nerve cells into a chemical sequence and then back again into nerve events faithfully reproducing the original item.

With memory one immediately associates the property of invariance, showing that we are accostumed to think in terms of a stable, storage-type memory consisting of enduring registers of engrams for everything that has concerned the brain in the past. In Sect. 1.2 we suggested another possible organization of memory, one that is essentially dynamic, and that renders largely useless the proposed enduring memory traces in matter. If we accept that the organization of brain memory needs not be represented by long-term, stable storage, then so-called memory registers or static memory should be of considerably less importance in man (if existent at all) than in other beings and perhaps limited to few reaction patterns related to particular sensory experiences. Even leaving aside these considerations, we feel justified in casting doubt also on memory's invariance on account of two well-established facts of everyday experience. On the one hand, man's brain needs to adapt its reactions finely to a wide gamut of situations which, on the other hand, correspond to widely changing operating conditions at the level of the individual neurons. Therefore an invariant response at each repetition of stimuli can hardly be expected from brain, since the associated conditions at the level of the individual neurons cannot be exactly repeatable to any degree of accuracy due to the convergence on them of inputs from various origins. The conclusion is, again, that a chemical-type, stably coded memory would be almost useless in man except perhaps in extremely unusual conditions, i.e., when the vast majority of other behavior-determining stimuli are suppressed, or their action is blocked in some way (if this can occur at all). Doubts about the invariance of memory-related behaviors also arise due to the fact that repetition is scarcely reconcilable with storage —better, repetition would render storage useless, and vice versa. It is commonly observed that a great many behaviors in animals, as well as in humans, require continuous reinforcement by repetition, so we may assume that supposedly predetermined, invariant behavior should be reinterpreted in most cases as being no more than the already suggested genetically facilitated acquisition of behavior or readiness of behavior exhibition.

Also that which is called instinctive behavior, that is, not learned or imprinted (admittedly the expression of species characteristics and therefore possibly genetically determined) may rest on sensory reinforcement and repetition of specific sensory experiences. This fact is easily observable in many so-called instinctive behaviors, such as those of the homing pigeons and of seals and birds retracing their reproduction grounds. Again, most wild animals seem to recognize their own territory less through comparison with a stored image of its geography than on the basis of their own odor left in it; continuous reinforcement is needed to keep the borders marked, for the "owner" itself as well as for intruders. However, the matter is not exhausted by these few considerations. Clearly, the possibility of laying down more or less labile traces in the brain structure is large; the real point is how labile they are, not as occasional and hardly repeatable events but as part of an organized and more complex system of traces of the same nature.

7.3 Memory and Eidetic Operation

Man's brain memory, whether the result of an automatic storage process or a painstaking eidetic operation, can always be retrieved, brought to consciousness, and introduced into reasoning. This means that reduction to eidetic form is the only way of using a memory. No one knows as yet how a memory trace is laid down, or whether it is the result of transformation of the elements provided by the senses or direct elaboration of reasoning itself. In either case a dual process of transformation of the item's elements is necessary: a) depositing the nerve impulse patterns describing the item in another form, the one appropriate for the nature of the "register" and b) retrieving the stored elements again in the form of nerve patterns. However, regardless of what the content of the memory is, either concerning images from sensory experience or ideas, their use in awareness requires that they have the form of eidetic objects.

This appears to be different in animals. For want of any evidence to the contrary, it is maintained in *PTN* that animals lack an autonomous intellectual function, i.e., the capability for spontaneously organizing ideas with a language of their own, something that may at least in a rudimentary way resemble autonomous human thought. Undoubtedly, every animal acquires knowledge of the world in which it lives, and this knowledge must find internal expression in a way suitable for use in the brain operations which give rise, not to reasoning, but directly to behavior. For the behavior to be consistent in repeated situations, the method of acquiring information must exhibit a certain constancy, and this has led to belief in stably laid-down memory traces. This view, however, means attributing also to animals the capability for organizing some type of abstract structures; although rudimentary, these would still resemble human eidetic objects. Perhaps in the past one has too hastily accepted the view that constancy requires a stable memory trace. As noted above, we should assume the existence of a stable memory only when the possibility of other, equally probable, and more simply conceivable organizations for ensuring constancy and repeatability of operations can be excluded.

In *PTN* it is maintained that animals, in place of autonomously organizing thoughts, can perform only associations. The associations which animals perform are those connecting the innumerable stimuli from the outer world with the limited number of categories of fundamental factors which regulate their life. These factors constitute the standardized categories of behavior that animals are able to exhibit. The following examples illustrate these: food and the acts associated with it, such as searching for prey or food, chasing, eating some and hiding the rest for later; danger, which is often associated with predator and flight from it or seeking shelter within the den; reproduction, with the theatrical enactment of various behaviors and later caring for offspring. The external stimuli merely need to be relayed to the triggering mechanisms of the elementary categories of behavior, the only items requiring stable memorization. Again, the latter does not have to consist of chemical sequences. In this case they could probably rely on genetic determination of the structure and organization of brain circuits, from which sequences of somatic and autonomic events originate, triggered in series one after another and each dependent on the preceding one as well as on sensory feedback from the surroundings. The relationships of every incoming stimulus to each of the categories of behavior can be recognized by automatic comparison of the characteristics of the stimulus with those of the categories, i.e., a matching in full or in part (as in the process of imprinting the counterpart of the effective stimulus to the desired behavior may be extremely rough).The involvement of awareness may hardly be of relevance in view of the fact that these fundamental categories of animal behavior are commonly linked to, and perhaps also dependent on autonomic and somatic reflexes and on secretions of neurally active molecules released directly within the organized structures by the stimuli normally associated with them.

The same may also occur in man, although such a dependence is surely not typical of man because his intellectual processes and his free will make him capable at any moment of interrupting the chain of behavioral sequences (which instead has an unavoidable course in animals). Thus, the cerebral function of highest degree in animals is represented by a structurally defined pattern of associations in which brain cell activity does not have the role of conscious promoter but that of stimulated executor, being urged to work by external or internal stimuli. While it is unconceivable that any intellectual process is involved in these operations by animals, this is generally the case in man. There is an enormous difference between man and animals in the time required to develop a correct response to various stimuli. Despite his enormously more complex and organized brain, man takes years to master the simple stimulus-response paradigm and gradually reaches full independence from it with the progressive ripening of his intellectual, i.e., eidetic, rational processes. No animal can learn how to construct the same eidetic objects as man does in responding to the same situations. Any behavior in animals that seems to contain a kernel of an eidetic nature (i.e., the rudiments of a rational decision) cannot be said to partake of a truly eidetic function (i.e., a dynamic, neuronally based process of reasoning involving consciousness-awareness, as in humans) but is as it were,"frozen," as in genetically determined neural connections, i.e., in genetically conditioned anatomical constraints, the only stable form of memory that we can image.

7.4 Human Brain Memory

Let us comment briefly on the conclusions of the preceding section. The highest functional stage of the animal cerebrum is reached with the ability to perform associations, however complex they may be. The interactions of the animal brain with its surroundings are restricted to sorting and distributing the innumerable stimuli provided by the outer or inner world —either simple or complex situations— to the relatively small number of structures controlling the categories of behavior of natural (and frequent) occurrence in their life. In principle, the criteria for distribution (the criteria for associations) depend on recognizing the congruence between outer items and behavior categories, that is, comparing characteristics of the item with those of the category. In animals, however, this may not be necessary. For want of autonomously generated rational thought, the procedure requires fundamentally an automatic organization (not necessarily a comparative procedure) outside voluntary, conscious control. In man the comparison is commonly guided by the eidetic processes and is therefore also a conscious operation. In animals the impossibility of constructing ideas or other kinds of autonomous mental representations means that no eidetic object can be stored in its memory. Ideas and the operations related to them simply cannot be contemplated in animals. Therefore, invariant categories of behavior must be genetically defined, that is, there must be a genetically defined association of classes of stimuli with behavior categories. Of course, it is not necessary for the eidetic meaning —inherent in every stimulus situation and behavior— to be genetically coded and for the congruence recognition to be performed on an eidetic representation of them. Much more simply and easily, the principle of association skips the comparison stage; the stimulus situation can elicit activity in (and the behavior category can directly and reflexly control) the same somatic and autonomic neural circuits and secretion of the same chemical mediators and hormones that are connected with the specific stimulus situations and with the corresponding standardized behaviors.

In man similar mechanisms are also at work in a number of stimulus situations and behavior categories; that is, in man there may also be genetically determined behaviors. The difference is that in animals these would represent the only processes of comparison (if ever present), whereas in man they are dominated by rationally conducted operations. In human brain, every operation is subjected to a judgement, autonomous, independent of external factors, and dependent solely on the principles that man imposes on himself.

We return now specifically to the question of human memory. We have said that man can be aware of his memories only in their eidetic form (since a memory is also stored in some way in matter), and that these must be reconverted to eidetic form to be used in the state of consciousness-awareness. Since the property of its invariance belongs to the concept of idea, memory as ideas and memory as registration appear merely to be two different ways of referring to the same thing, i.e., objects to be retained. But, this question cannot be dismissed as a mere choice between different terms. What ever it is that an idea represents is somehow memorized. The idea itself, therefore, *is* a memory in some sense, representing the stable form of a concept, and the phase space of all our ideas thus constitutes our overall memory. Consequently, the way in which the brain conceives ideas in its neural circuits likewise outlines the structure of memory. However, memory in the sense of physical registration implies the property of being materially defined and permanently localized in space —properties that are altogether foreign to the concept of idea even when we attribute to it the property of invariance. Somehow, it seems incorrect to identify idea too closely with memory, particularly because that in the brain which is materially defined may also be genetically defined. If this is the case, a large part of our thoughts and our reasonings may have to be seen as genetically conditioned —which would be an unpropitious conclusion.

Etymologically, idea means "what appears," thus we use the term in the sense of the form or the representation, not as expressed in matter but in language, either in an internal code or in the external expression. An idea is not something physically tangible, and consequently the property of invariance hardly seems appropriate to it, if invariance is to be associated exclusively with stability of structure, that is, structured matter. The concepts of idea and memory must clearly be better differentiated. The principal issue here concerns invariance, an essential property of every memory and every idea. The extent to which absolute invariance is an essential character is debatable in the case of both brain memory and abstract structures such as the ideas. A representation that is sufficiently stable to allow a reasonably faithful reproduction on repeated occasions of retrieval, cannot be expected of brain memory. Even if all the vast body of information arriving in the brain could be coded in sequences of molecules deposited somewhere in its structure, absolute stability would be impossible at least in the long term, unless these were structurally organized according to the principles of reproducible genomes. Instability would result from molecular turnover, interference of metabolic needs in the chemical processes of cells, and cellular decay itself. Indeed, if one examines the degree of stability of human memories, one discovers rather less invariance than may generally be assumed. Even the memories of recent events may vanish quickly; if not, repeated retrievals of these show progressive reduction in details and therefore a reduction in the amount of information contained in them. If there is any part of the memory that remains invariant for some time, it is the essential core of meaning, although even this may lack invariance in the long term (years).

Regarding ideas we may advance analogous observations —in this context let us call them memory ideas inasmuch as we are investigating their property of being at the same time also memories. Rather than taking for granted that the memory ideas in general are invariant, we should actually wonder at the fact that they exhibit any invariance at all. If these are coded in matter, they would either be forgotten or would undergo changes in content (such as simplification) as every other item similarly stored. If not stored in the brain hardware, the memory ideas could exist in the brain only as "fleeting" objects, abstract in nature but unavoidably of some neurological nature (e.g., nerve impulses). Ideas are part of intellect and knowledge, but their neurological feature makes them potentially subject to all the manipulations that the eidetic processes are likely to introduce into their content when stimulated by those factors —other events, information, reasoning— that normally modify, interfere with, or in some way condition the eidetic processes. However, despite the possibility of undergoing changes between successive representations, the concept of idea entails that each time the memory idea remains "true," in the sense that every time it tells its "present" truth (no matter what the previous truth was, since this would be totally lost, i.e., forgotten).

There is something very disturbing in the preceding paragraph, because it suggests the possibility that we may have no way (at least in several cases) of determining whether the memory idea is true or false. For instance, when it relates to a temporally, logically ordered train of ideas concerning a past event. If, in a memory idea a strong interaction with foreign eidetic material at some point in the same circuits has reached the core, i.e., the essential meaning, this may not retain its former content, which means that the memory idea may cease to be the same. However, since the sole memory idea present concerning that object has changed (and no "copy" of the former one exists), the changed memory idea, although false, can only be considered by mind as true. However, there may be even more in this. An apparent memory idea of something that never existed may be evoked if suitable stimuli are delivered to cerebral structures (as occurs in neurosurgery), autonomous or induced representations of some situation (particularly when loaded with emotional content), provided they reach the eidetic structures and succeed in setting them in motion. For the mind that "created" it, the memory idea nevertheless is true (even if perhaps only briefly); it is real because, once created, nothing can distinguish it from other memory ideas related to something that really did exist. Thus, the content of memory ideas in our mind has a priority as "really true" over those of others; we first trust "our" memory. Some human psychotic phenomena, results of cerebral stimulations, hallucinations, sensations of déjà-vu, some types of epileptic aura, etc. may perhaps be explained in this way. A large amount of evidence points to the fact that in every individual abnormal sensory experiences, such as illusions, or true hallucinations, can easily be induced by drugs or the stimulation of certain structures. The links between these artificially induced phenomena and true eidetic objects, such as memory ideas, must yet be made clear. (Notice that no distinction has been introduced here between memory ideas concerning events, consisting essentially of a specific ordering —temporal *and* logical— of the individual component ideas, and memory ideas consisting of one individual idea. The point is that both types of memory ideas are made up of elements connected exclusively by a logic order.)

None of the properties of memory ideas discussed in this section is shared by animals. We maitained in *PTN* that the fundamental difference between animal and human brains is the absence in the former of eidetic processes, due to the lack of an internal language —no clear-cut, unequivocal trace of it having been detected so far. There may still exist some rawer type of animal eidetic at the level of memory, on the one hand, and at the level of recognizing the meaning of behavior-evoking stimulus situations, on the other. As should be evident by now, in our opinion the two questions are closely linked —in fact, two sides of the same problem. Memory, in itself, can be conceived only as an ensemble of memory ideas since every representation in the human mind is such by virtue of a language, and man, whatever he may want to memorize, does it in terms of items translated into memory ideas (for instance, house, street, relatives, dog, car, trees, hill, see, food, bad, good, pain, moral, religion, science, philosophy, society, politics, etc., together with all their related and dependent ideas). On the other hand, for every stimulus situation, mind is forced first to recognize in its own logical terms the nature and meaning of the events and then its own relationships to the stimulus situation (to be able to organize behavior). Therefore, there is no objection to seeing the necessarily resulting behavior as being dependent on fully automatized mechanisms in animals and on fully rationalized ones in humans.

7.5 Eidetic and Noneidetic Memory

It is quite reasonable to assume that in animal brains no eidetic processes take place, although this view is based exclusively on negative evidence, i.e., on the lack of unequivocal signs of autonomous eidetic activity. Before accepting this view, however, we should explore certain of its implications. When the absence of language precludes eidetic activity, memory can be present only in the form of noneidetic representation, and every hypothetical process of recognition then depends on some automatic, noneidetic mechanism. These two points need further comment, since this position suggests the existence of two types of memory.

Eidetic memory can result only by transforming an idea from expression in language form into another form. However, the transformation must be such as to allow a topological, isomorphic transposition in both directions, into and out of memory. It is unimportant whether memory consists solely of nerve impulses, patterns of closed and open synapses, or sequences of molecules in the cells, provided that we have the code and the physical means by which language can be effectively and quickly transformed, or transposed, into memory in a thoroughly reversible way. Possessing that code, one could "read" the memorized items —ideas, either isolated or in sequences— i.e., repeat sentences or representations of events, images, etc. On the other hand, noneidetic memory could not consist of a transformation of eidetic objects, since the latter, being actual language, would retain after the transformation the character of eidetic-type memory. Therefore, noneidetic memory can only be concerned with items which the brain does not (cannot) translate into language. In humans these are only the items that the brain cannot know (since knowledge is eidetic) such as the characteristics of the body and the general properties of its own structure, which appropriately take the form of genetic memory. However, in animals, although lacking in eidetic capability, noneidetic memory apparently can also not be of unlimited use because many vitally important behaviors are based on imprinting after birth; this procedure strongly resembles memorization, but it can be interpreted much more easily as a facilitated association of fixed stimulus-response paradigms in anatomically determined structures supporting an organization akin to reflex arcs. In animal life, the quick adaptation to new situations would be more easily and efficiently achieved through the interaction of numerous structures organized in a reflexlike way than through the new rearrangement of molecular sequences of a memory laid down in matter. The reflex arcs may compete, each respective with its weight, in determining the final behavior, with the enormous advantage of allowing continuous, automatic, and fine adjustment to the contingent variations occurring in the surroundings, within a wider framework of organized reflex reactions.

I am firmly convinced that the indiscriminate use (actually abuse) of the term memory has introduced a lamentable confusion. The concept that where there is order of some kind, there is also information, has led to the common view of the identity of information and memory (as stored information). The term memory is generally taken to include information stored in a somewhat stable and retrievable form. This blurs the thermodynamic meaning of information and of its opposite, entropy (having surreptitiously introduced the foreign property of stability and reversibility of mechanisms), and it has no doubt also been the source of substantial misconceptions. Using a single term for the so-called genetic (or molecular) and "neural" memories is not yet justified, and no hypothesis is able cogently to unify the genetically and the neurally based mechanisms. We should not consider genetic determination as any memory at all; genetics merely determines most structural and some functional properties —those strictly inherent in the physical organization of the structure. However, human memory, as it is commonly seen in neurosciences, is concerned only with eidetic processes, i.e., it is a language-related function, and as such it is concerned with maintaining logical procedures and abstract objects at the same time. The logic that underlies rationality and thus eidetic activity cannot be altered, and therefore it cannot be determined to any extent by the genetic mechanisms that determine the body characteristics. The eidetic processes make use of the genetically determined brain structure only as a physical base. Concept-ideas by their nature and dynamics are subject to formulation, development, and reshaping in a way that is completely foreign to the logic which supersedes the principle and the mechanisms of genetic determination.

Thus, let us say that in animals there is no memory at all, since what is produced by their brain as stable patterns of behavior is either genetically determined or the product of "associations" among activities of neural structures, stabilized earlier or later in life. In animals, the relationships between what may be called learning and brain memory are still much debated, especially because of what learning would imply in terms of eidetic activity when no autonomous eidetic processes seem to occur to any extent. As we noted above, in animals any stimulus situation must match with one of the fundamental, genetically determined categories of behavior, that is, it must be recognized as congruent with or related to one of them (and this can occur in a fully automatized way), or the stimulus is relayed directly to the behavior-triggering mechanisms. The problem in humans is more complex and shows an organization based on quite different principles. In animals the problem of memory cannot exist since the problems are solved by neural mechanisms that automatically assess congruence or more directly by a network of anatomical connections in the structure. In the case of man the problem can only be solved in terms of memory as such. Of course, man may also occasionally exhibit standardized behavior, such as flight from danger and sudden aggressivity, but in man these do not show the unavoidability that they do in animals (see "Appendix" to *PTN*). Therefore, we can also see in man the automatic coupling of basic stimulus situations with fundamental categories of standardized behaviors, the latter, however, are integrated into a much wider and more complex structural organization of the brain (due to philogenetic and ontogenetic development) which makes man's behavior essentially free from any rigid conditioning.

7.5.1 Memory and Mental Representations

To conclude our observations, we distinguish "autonomous" human eidetic activity from the "nonautonomous" animal form. One must ask, actually, whether a process can be termed eidetic if it is not autonomous —if not as the exclusive, at least as the most relevant of its manifestations. Autonomous eidetic activity refers to thought that is independent of recent and remote causes, whether internal or external. Thus, only free will can be the basis of autonomous eidetic activity (see also Sect. 10.3). The contemplation of abstract objects that are not (directly or immediately) related to material reality reveals the autonomy of the operation, because thought can originate here only as an independent drive, forcing mind to create abstract objects and to develop and transform them. There can be no doubt that man has the capacity for such an independent, free, and autonomous eidetic function. This is manifested in all his characteristic behaviors, achievements, and products of mind —science, philosophy, art— in other words, in his capacity for abstracting.

What kind of knowledge does an animal have of the world around him? It surely reaches some kind of knowledge of that which immediately concerns its life, but under what form? Satisfactory answers cannot yet be provided to such questions because they would necessarily rest on presumptions for which, as we said above, we can offer only negative evidence, i.e., that no animal has so far exhibited signs of true eidetic activity. Many of today's species inhabited the earth long before man, but the brain of none has developed beyond its initial stage. I believe, as repeatedly stated, that in animals the lack of an "internal" language manifests itself in the failure to develop the brain's structural organization supporting an "external" language suitable for reasoning, as occurred in humans. This negative evidence still does not rule out the possibility that in some obscure, indefinite, or not clearly definable way the animals also possess some kind of concepts of their own identity, the conditions relevant for the self, or the immediately surrounding world. However, since these cannot be expressed by the animal brain in language, they could only resemble "pictures" of single items or of composite ensembles. Thus, a picture that is described by the human mind through language and eidetically examined in its component exists in the animal consciousness in a quite different way; this may be as a mere "presence" without a definite shape because the lack of eidetic function prevents one, and yet there is a potential for behavior-driving power. This is not mere speculation. On the one hand, in man the two cerebral hemispheres also manipulate the concepts with rather clearly differentiated procedures; this is more language-symbolic on the left side and "spatial" on the right side (a representation still requiring definition, though perhaps no less complete than the former). On the other hand, mammal brains, particularly those of antropomorphic apes, have developed much farther than those of other animals; besides a morphological differentiation they have also considerably expanded in overall volume. Such an expansion entails an increased number and a much greater complexity of neuronal circuits. Although not yet as developed as in man's brain, the ape's brain may subserve fundamentally the same function of elaborating a "spatial" picture, albeit somewhat raw and hazy.

These considerations suggest rather less of a gap between the capacities of man's brain and those of animal brains. Below, we examine the possible organization of eidetic objects —the concept-ideas— at the neuronal level; there we shall find that the elementary circuits may be rather simple, and that animals perhaps also possess these. In this case eidetic activity could not in principle be altogether ruled out in animals. In addition to neurophysiological issues, this would certainly also involve metaphysical considerations. However, comparative knowledge of the conditions of structural organization in the two types of brain are still far too insufficient to consider such issues at the present time.

8 Ontogenesis of Ideas

In examining the ontogenesis of ideas, we consider two types of ideas, the natural and the constructed, which differ as to type of constituting elements and symbolic representation. Not as a corollary of this but as a fundamental difference between these types of ideas, voluntary eidetic motion can be seen as the primary source of constructed ideas, sustained by an eidetic operator. The discriminating factor revealed through the latter resides in the symmetry or asymmetry of relationships between the operator and the idea, depending both on the nature of the elements and on the operator. Neither operator nor idea can be considered as "isolated" in thought but are always in the "complex operator-idea," and the type of idea is defined by the properties of the complex. A particular type of operator, however, is the "time operator;" this produces effects over long periods of time and thus causes the ideas also to grow older; the mind thereby becomes wiser and more mature. In this process, abstraction progressively penetrates the hidden, essential content of the idea. As an operator of abstraction, this leads to identification of essences; the time operator is an extremely composite object, consisting of all the knowledge accumulated as part of mind's horizon, the horizon of all the ideas, and relates the essences to one another. Through this procedure, associating one idea with all the others, each idea can be progressively refined and reshaped until its essence stands out at its clearest; this is the true process of acquiring knowledge.

8.1 Natural and Constructed Ideas

Since human memory is eidetic memory, i.e., it consists of ideas, we are led finally to the topic of ideas as "objects" of memory and thought. We outline, first, the heuristics of idea formation. We have been regarding ideas as "abstract" structures; we add now that an idea is made up of sets of elements inherently congruent by nature and held together by a coherence intrinsic in the specific content of each idea. An idea defined by such an intrinsic coherence constitutes an indivisible unity. Furthermore, the idea is an independent, self-sufficient entity, for it finds the reason for its own being in an inexpressed and inexpressible "concept" and therefore in itself as the "form" of that concept. As a consequence, once created, the idea becomes also self-sufficient in relation to the brain which conceives it, in the sense that it is the representative of an essence, of a "universal." In fact, mind can only be clearly aware of the "presence" in itself of a concept, an abstract object of still uncertain structure, and yet bearing a significance which offers itself to immediate understanding as an apodictic reality. Mind gives this concept the form of universals —the ideas— through those processes that we say, rather incorrectly, conceive ideas, i.e., the eidetic processes.

We must recognize here the existence of different types of ideas, differentiated by the degree of affinity in the constituting elements. For an idea to result from them, the elements of an idea must possess an inherent congruence of nature linking them to one another as members of a class; they must also be consistent with the content of that idea, i.e., there must be an intrinsic coherence among the elements making them specific to that idea. However, the elements may exhibit different degrees of congruence among themselves, in the sense that the affinities linking them to one other may be more or less clear-cut, independent of the content of the idea. In this way, the ideas are built upon a double basis: on the one hand, the reciprocal congruence of their constituting elements, inherent in the latters' nature, and, on the other, the intrinsic coherence binding together those particular elements to form an idea with its particular significance. This double basis is typical of ideas representing objects of the physical world, but it is not at all confined to the latter. Certainly, in the physical world one finds analogies for understanding the nature of other, non-matter-related ideas. For instance, a chemical reaction can be obtained with elements having a chemical nature which alone defines the reciprocal congruence inherent in them. In turn, every specific reaction requires that the elements must also be those "specific" elements (of different chemical species) alone entitled to enter that reaction —their specificity for the reaction resting on the fact that they are the only ones possessing the intrinsic coherence defined by, or dependent on, that reaction. Because of their frequency in the world of nature, we can call the ideas having these characteristics "natural." However, also in the immaterial, metaphysical domain there are ideas with the same double basis, for instance, the idea "good." A great many elements may contribute to the general idea of good, all having in common the same nature, that of being elements of good. The latter, however, may assume the most diverse features, but for a given, specific good, the elements are sorted out by an intrinsic coherence proper only to the idea of "that" good.

For another type of idea the inherent congruence among the elements either does not exist at all or is not as clearly apparent, as in the former type. Thus, these ideas are founded on "disparate" elements, and an intrinsic coherence among them —the unifying link— is only imposed by the content of the idea. Although still imprecise, the term "constructed" seems sufficiently appropriate for ideas of this type since mind's effort to recognize an intrinsic coherence among elements (sometimes widely different) is an autonomous attempt at constructing something new, something not previously existing —thus an act of creation. Quite differently from the ideas of the first type, the constructed ideas are generally the result of reasoning, and the place of each element in relation to the others is therefore weighed, compared, and debated until the new intrinsic coherence is established on logical bases; only then does the idea begin to exist. Of course, the coherence among the elements of a constructed idea must at least potentially exist and it is up to mind to recognize this. However, the exact relationships binding these elements in a coherent unity are not initially necessary, exclusive, or indubitable; they become so only after the logical links for that specific content of the idea have been uncovered. To the constructed type of idea belong many advanced ideas of science and most ideas in philosophical, ethical, social, political, economical, and religious areas.

However, the difference between natural and constructed ideas is greater than appears at first sight. It is not simply that the former are founded on generally inherent congruence among the elements and on intrinsic coherence of the specific elements, whereas the latter result from coherence of noncongruent elements only. The fact is that the lack of an immediately recognizable, inherent congruence among the elements actually hampers a clear-cut definition of the whole set of elements that *might* belong to the constructed idea. For this reason, the advanced scientific theories based on constructed ideas and all the philosophical, ethical, religious, social, and economic theories that are almost exclusively based on constructed ideas are never free from some degree of indefiniteness and may therefore be submitted to reexamination and discussion over and over. This produces almost infinite shades of content when the ideas comprise many common elements and differ in only one or a few. However, there is even more to this. As discussed in Sects. 4.5 and 5.3, many constructed ideas are actual field-ideas or even principle-ideas; because of the high degree of indeterminacy in their constituting elements they represent many fields and principles with unknowable boundaries.

At this stage, the point is not to provide further definitions of types of ideas. By introducing parameters other than congruence and coherence certainly the types may be better differentiated, and with the definition given here the assignment of many ideas to a particular type can always be questioned on an epistemological basis. The point, rather, is that in the construction of ideas, will is the essential factor, for mind directs the procedure in a desired direction with the explicit purpose of finding the rational relationships linking sets of disparate elements. Since the implied rationale may not be clear beforehand, will continuously searches for the clearest possible definition fitting an aspect of the underlying concept, i.e., a new idea. The search itself implies that at every step in the reasoning —selecting the elements of the idea— will performs the choice after comparison of alternatives, assessing the coherence by rational arguments. However, preceding the laborious and painstaking definition of the idea, a more schematic picture of it, a faint outline —besides the isolated initiating concept— must be present from the beginning even if in such a form that it still retains many obscure aspects; otherwise the starting point of the procedure for constructing the idea would be left completely to chance. The sought-for coherence may be present initially in an internal language of still unknown structure (probably not the same as the spoken and written language; see Sect. 10.6). For these reasons, it becomes absolutely necessary to translate, or transpose, the element-ideas onto symbolic, spoken language —i.e., onto the least possible "spatial" or "figurative" representation— the only language allowing the weighing and rational discussion of the arguments for each of them.

8.2 Implicit Ideas

The exact nature of the parameters for compariing the elements' contents, their selection and the recognition of the idea's unitary essence remains a matter of guesswork. Above, we likened the idea, on the one hand, to the representation of a physical law (Sect. 1.4) and, on the other, the assembling of such idea-law in the neural circuits to the implementation of implicit functions in analogue computers (Sect. 1.7). No arguments have been advanced to justify this line of thought other than that it appeared sufficiently reasonable for exploration. It is possible now to return to this and to give a more complete description with the arguments advanced in discussing knowledge, space, mind, and memory. We can begin by founding on them a unified theory of eidetic activity consistent with the more general neurophysiological principles explored in *PTN*. It may seem far-fetched to liken the assembling of an idea, through the gathering of its constituting elements, to the implementation of implicit functions in analogue computers. However, there is something so strongly appealing in this principle of implementation, as we shall see below, that I whould posit it as an hypothesis to be tested. In point of fact, the hypothesis includes the identity of the idea's abstract structure and of its physical counterpart, in the sense of the existence of a single logical principle defining both the way in which the idea takes on its abstract shape, on the one hand, and its physically embodied shape in the neural circuits, on the other.

The picture that we have produced of the idea is that of an abstract construction (an abstract object) which in order to exist must be embodied in the matter of the nervous system and in such a way as to faithfully represent the idea's fundamental content. A number of characteristics must be accounted for in the process. a) The idea represents a system of relationships among the constituting elements; likewise, the formulation of a physical law represents the relationships connecting the physical parameters. b) Each idea, although comprising several elements, is in essence an indivisible unity. c) The appearence of the idea in mind in a state of awareness may be gradual or sudden. d) Each idea, in addition to representing an invariant in terms of its essential content (the kernel content), may simultaneously present multiple facets and shades of significance. e) An idea may constitute a "field," i.e., a continuum of tightly connected ideas with the essential kernel in the idea core and through this field may influence the eidetic processes. The same representation provides the basis for connecting the eidetic process, as embodied in the brain, with memory. Let us try now to identify the neurophysiological elements that support the hypothesis of a common, connected complex of the fundamental and manifold aspects of ideas and memory.

First, we should recall again that the elements of an idea need not be simple, elementary unities in themselves, but that they are more frequently represented by other ideas, and although still indivisible unities as far as the meaning is concerned, they comprise clearly differentiated parts. All the necessary elements must be present simultaneously, sometimes as part of other ideas or, more generally, as ideas part of a reasoning. The fact that we are able to represent to ourselves isolated elements, suggests the logical identity of the "substance" of the elements and of the ideas in general as invariant entities —as so many abstract bricks in the eidetic constructions. Therefore, we must wonder about the nature of the agent which integrate the elements into a new abstract construction, a new idea. (Above, we ruledout that one can be confronted with the problem of recalling from a registered memory, since we doubt even the existence of the latter.)

As we have said, a possible answer to our questions —conceptually simple and yet explaining many of the complex phenomena related to eidetic processes and to memory— may be found in the principle of implicit implementation of functions, on the analogy of that which generally operates in analogue computers. Here the fundamental circuit is a feedback comprising a suitable number of operational amplifiers, where the variables of the function constitute the inputs of the amplifiers. The principle of the implicit implementing operation provides that most of the variables are created in the loop itself, led off at some step of the operation, and fed back as input to a summing amplifier. This means that not all the variables need to be present initially but may be "constructed" appropriately in the course of the procedure. For the well-known characteristics of the feedback loops, after the short time elapsed for the currents to reach regimen, there is no further reference to a "beginning" of the operation. In fact, once the variables entering the general equation implicitly as $f(x,y,z,..n)=0$ have been made suitably to interact, the function physically "exists" and remains as long as the circuit remains closed and the inputs active. Of course, the circuits made up of neurons cannot be likened generally to operational amplifiers circuits; only ideally may we transfer the principle of implicit function implementation to the implementation of ideas, i.e., to their physical embodiment in neuronal circuits. Let us see which of the properties of these feedback loops operations fit the concept of idea. The idea results from the assembling of elements, sometimes reciprocally congruent and sometimes not, but once they are allowed to interact, the idea exists and it remains as an indivisible unity. The basis of the hypothesis is that in both computer and brain the idea is assembled in a feedback loop, a structure ensuring the necessary unity already immanent in the generating principle, that of a function based upon an intrinsic coherence in the ensemble of constituting elements (in the case of the computer, the intrinsic coherence resides materially in the links between the elements). To avoid difference of semantics with the arguments used in likening the idea to a physical law, we may say, more naturally, that the idea is a function of its variables, thus identifying it with the elements entering the definition of the idea.

In operational amplifier circuits, further feedback loops may be variously coupled to represent (to implement) more complex functions. This feature —common in the the structural organization of the central nervous system— corresponds to the fact that neural feedback loops can be inserted into other loops, thus ensuring that some of the elements of an idea are represented by other ideas, while all the elements still maintain their nature of self-sufficient entities. Moreover, the progressive development of the ideas is achieved, in turn, by simply introducing into, or subtracting from the loop other elements (other variables, which may also be represented by other loops); this necessarily changes the content of the original idea since this is always a function of the variables present and operating. In the case of implicit implementation in operational amplifiers, the onset of the implicit function may be flashingly quick or slowly building-up, according to the rising velocities allowed in the involved currents. The same has been said to occur for the onset of ideas in mind, although here the time involved may obviously be much longer —this is the time of the neural operations necessary for recruiting the elements, particularly when these are themselves other ideas, and for making them to interact. However, once the interactions have reached their regimen, there is nothing against the "flashing" existence of the idea, and its remaining existence reveals its invariance (as invariant as the function implemented in a physical circuit, provided the circuit maintains that particular structure, and the inputs remain active).

The change and development of the ideas by addition, subtraction, or substitution of variables indicate that the invariance of each idea is founded only upon invariance of its elements. Since as far as we understand, the elements are represented in the nervous circuits by nerve impulses and cell transmembrane currents, i.e., by events of very limited duration (milliseconds), an idea can be maintained unaltered in the nervous loop only by continuously feeding into it the suitable input impulses (or currents). While this might appear at first sight to constitute the greatest menace to the invariance of the idea, it actually corresponds to the essence of its abstract structure, an incorporeal and fleeting entity always present because (or if) always renewed; (the double aspect of the idea, as an incorporeal, abstract entity and as a physically existing one is discussed in Part V). Thus, the developments wich an idea undergoes as its elements progressively change correspond to the possibility that some of the inputs are either casually or voluntarily changed (e.g., by activating or suppressing some of the variables). Until the essential (kernel) content of the idea remains unchanged, the idea (regardless of whether natural or constructed) can be seen as building around itself shells of closely connected contents, i.e., a field originates from it as in a continuum (since at each point of the interrelated feedbacks there remains something of the original idea). Considering that many elements may be shared by different ideas, through these elements the ideas may also become congruent with other ones, thus exerting their fundamental function in thought development and becoming to it. In this way, the implicitly implemented functions in physical circuits, by interacting each other, that is, by suitably coupling different functions, fulfill the purpose of building larger, articulated, organized functional complexes.

According to this view, the eidetic system organization consists of a maze of feedback circuits variously interrelated, very rarely simple and generally quite complex because the neurons of the central nervous system are interconnected in an extremely widely encompassing way, and therefore the inputs to each feedback circuit are commonly multiple. Potentially, each input may change the state of a circuit in at least two ways: a) by closing an open feedback through the excitation of a key neuron or by opening a closed circuit through inhibitory synapses and b) by introducing into the circuit a new element, thus inducing a change in the previously implemented idea. Since the origin of the elements composing the idea are most varied, each element must be submitted to a judgement on congruence-coherence, and ideas are then built up of the "approved" elements. The intervention of will (and thus of consciousness) is not always required, at least not for the natural ideas (in contrast to constructed ones), and this makes it easier to understand how the same elements can be used independently in different ideas simultaneously present, each of them resulting from the assembly of many such elementary circuits, including commonly shared ones (see Part V for a discussion of the relationships between ideas and state of consciousness).

The fundamental characteristics of the idea as outlined here are ensured by a principle of embodiment in the nervous structure similar to the principle of implicit function implementation in analogue computers. The extremely different physical structure of the cell and fiber circuits of brain from the operational amplifier circuits of computers, does not contradict the striking similarity of their respective properties. The neural multiple-input feedback circuit comprising many neurons is in fact functionally even much more flexible a structure than its computer counterpart since different operations can be performed at each step (perhaps, at each neuron), depending not only on the position of the neurons inside the circuit but also on the pattern of the activated synapses. If the equations representing the ideas were not to be implicitly implemented, such mathematical functions as derivatives of every degree, integral, ratio, etc. would have to be available before their introduction into the circuit implementing the equation and therefore not obtainable in the same circuit but "explicitly" produced in separate circuits. The latter procedure is often a necessity in analogue computers in the case of nonlinear functions (such as powers, roots, etc.) and represents a significant limitation due to the size of the instrument. Certainly, this does not occur in the central nervous system, for the neurons composing the circuits are typically nonlinear in nature, thus simplifying the necessary operations within the same circuit. However, the important point is that in the closed configuration of the feedback circuit lies the unity of function and concept —that is, of meaning— inherent in the nature of the idea, exhibiting the properties of abstractness and physicality of a single structure that is necessary to exist in the brain. At present there is no reason to reject this hypothesis about the neural status of ideas, an hypothesis which in view of the wide feedback arrangement throughout the brain seems most natural. These are not by themselves decisive arguments, but the extreme conceptual and structural simplicity in this view of the implicit representation reinforces their strength.

8.2.1 Examples

So that our arguments do not remain but formally superficial, if interesting, analogies, it is necessary to provide experimental confirmation for the fundamental point thus far taken almost for granted, namely that the ideas can be represented as mathematical functions, allowed to be implicitly implemented in the neuronal circuit. The essential point is whether the ideas are *always* representable as implicit functions. The answer is affirmative *if* the ideas are representable in the form of an equation, with the elements of the idea in the reciprocal relation of the general form $f(x,y,z,..n)=0$. While this may appear as a far too rigid and perhaps unpalatable proposition, we have only to think that a symbolized mathematical formulation is simply a shorthand way of expressing a logical dependency, i.e., the logical relationships among elements or among ideas. For this reason, each element must reach the necessary degree of abstraction in relation to the overall significance of the idea in order to fit in as an element of the general equation. At least for ideas strictly related to material objects —natural ideas— the procedure may be simpler than this. Here it would be extremely easy to establish also quantitative relationships (for instance, ratios) among the elements regardless of the units of measurement, even arbitrary ones.

Let us consider as an example the idea "container." The elements possibly entering in this idea are a) solid object (or volume object), b) internal cavity, and c) walls. (Note that all the elements are themselves ideas, which in turn require other elements for their definition.) The corresponding terms of the variables are x,y,z. The equation $f(x,y,z)=0$, may be solved in any suitable permutation of the terms (to be embodied in feedback loop circuit), for instance, $y+z-x$. The concept of container could also result from the ratios x/y, x/z, and y/z. The condition $y>z$, could be set to a value much higher than that corresponding to the threshold of transition between the simple "solid body with a hole" and "container." The latter two ideas are different, though strictly related, since they are made up of the same elements and are likely to coincide at an intermediate, ill-defined meaning for the smaller values of the ratio y/z. (The general form for the implicit representation, in operational amplifiers feedback circuits, of ratios such as $w=y/z$, would be $wz-y=0$. However, as far as the neural circuits are concerned, we cannot determine the best representation of a ratio, although the required neuronal operations may be sufficiently simple.)

Within the same schemes, ideas such as night and day, and all the other concepts based on the ratios between elements, could be likewise obtained. We can also offer other examples. Some ideas can be obtained through logical operations of the "derivative-integral" type, having the general nature of transformation from small scale to larger scale (or vice versa), that is, from particular to comprehensive, such as man-mankind, through successive steps of higher aggregation. It is typical of the analogous implementation of these functions in implicit form to extract the individual ideas (functions) stepwise through stepwise integration starting from the higher derivatives. (The implementation of partial derivatives is not as straightforward as for time derivatives, although, again, the procedures in the nervous system may be simpler than in operational circuits.) These examples may suffice to give an outline of the heuristic procedures involved. However, if the possibilities were limited to natural ideas, the hypothesis would be of no use. It should be possible through suitable analysis of the elements to extend the principle of symbolic representation in equation form to complex ideas, such as those pertaining to philosophy, ethics, etc. The problem cannot be tackled directly here because the ideas typical of these regions are generally principle-ideas which need first to be analyzed (divided) into their constituting element-ideas, each of these first steps representing another epistemological effort. However, once the component element-ideas have been isolated, there should be no difficulty, in principle, in describing their relationships in symbolic form, as one commonly does with physical laws even of substantial complexity.

8.2.2 Implicit Implementation and Memory

The view of the idea as a neural function implicitly implemented in the neuronal circuit suggests several points concerning the question of "neural" memory. As discussed in Sect. 7.2, the idea's content (its meaning) has a kernel of invariance maintained on repeated recalling even after long periods of time. The obvious explanation is that the ideas are registered in memory (whatever this may mean). Present theories about the mechanisms of memory still center on two main categories: a) a long-term memory, based either on a stable fixing of the item into some physical medium or on stable physical arrangement of circuitry; and b) a very short term memory, based on an indefinite (and I should say, very questionable) ability of the nerve messages to recirculate within neural circuits without suffering loss of informational content. At the present, either category meets with considerable opposition on various grounds.

The heuristic procedure of implicit implementation assumes that the suitable inputs to a circuit are available, and that after the short time fraction for the inputs to reach their final value the ideas exist and remain if the inputs remain active. This, in a sense, is a way of memorizing items, not by physically storing corresponding engrams (which has until now not been demonstrated) nor by allowing their neural counterpart (patterns of nerve impulses) to recirculate freely (which is not demonstrable, and at any rate would entail the unavoidable and dramatic loss of information within seconds). Rather, memorization is achieved by keeping the corresponding neural implementation of the idea continuously alive, a feat made possible only if the inputs themselves remain alive. In this hypothesis, the relationships between eidetic constructions and memory are viewed from a common, new perspective. According to this picture, ideas and memory are made up of the same "substance," both are fleeting, abstract structures, taking on life every time the inputs are activated and for as long as the circuit's components (cells and connections) are allowed to interact with each other. The common essence of ideas and memory lies in the system of relationships among the component elements, thus not in matter in itself but in its fleeting patterns of organization. There is thus no difference between ideas and memories, neither theoretically (abstractly) nor in their concrete existence. In this view, invariance of idea and of memory is provided by the invariant nature of the relationships among the constituent elements of the idea, even when the elements are themselves ideas, since their elements are invariant. The relationships among elements, in turn, are represented in the neural connections through which the nerve impulses travel, established at the moment of formation of the idea. This abstract and simultaneously physical and physiological construction can also allow changes, i.e., produce more or less different ideas and memories when other inputs are allowed to activate the circuit, as well as returning to the previous content (meaning), when the constellation of inputs returns to the former configuration.

The extreme flexibility of this hypothetical construction is emphasized by the fact that by itself it does not contain any necessary prescription concerning the space allocation of the memory-ideas. For greater clarity, let us think of an artificial system composed of interrelated nets of feedback-arranged, neuronlike elements. In this system there is no need for localization of eidetic or memory structures; any memory-idea can appear wherever the circuits are appropriately connected with the necessary inputs. This does not appear at first sight to be the case with the nervous system. However, here too, since its circuits consist essentially of feedback loops variously interrelated and widely spread throughout matter, memory-ideas could arise almost anywhere, requiring only that the inputs be present and fed into the necessary connections (within the loops). In other words, in the nervous system, assuming that the circuits are widely diffused throughout the brain, a strict localization of memory-ideas would be determined only by a more circumscribed distribution of the inputs. We should remember, however, that even in the case of natural ideas the inputs are hardly fed directly from sense organs (a fact that could allow only the assumption of spatially circumscribed influence) but from structures responsible for more advanced stages of sensory elaboration; therefore, the inputs might reach the eidetic circuits from wide and distant areas. Not every feedback loop, especially the very simple ones, needs to be contained within a small volume of substance but may connect cells of different brain regions, and the inputs may reach these loops in distant areas. These two facts mean that the idea and the corresponding memory representing it may be possibly "dispersed" over wide regions of the brain without losing any of its properties. Moreover, the presumably ubiquitous character of the basic structural organization supporting the ideas —the loop— certainly facilitates the formation of very complex unities made up of element-ideas and memories of different nature and pertaining to different regions of knowledge.

This hypothesis shifts the weight from the identification of very specialized structures for producing ideas to the organized activation of the inputs to circuits, rather simple in principle but intricately arranged. The next obvious question, then, is how an idea may be "recalled," since the problem is now to understand where the necessary inputs are activated. Again, the generally accepted answer would be that recall comes from memory, because no one can see where inputs might originate if not from some memorized item, either word or image, which activates the appropriate constellation of inputs to the eidetic circuits. Of course, recalling through symbols (words) cannot be viewed as a true memory recall since symbols are language, and this is not yet sufficiently and clearly connected with a memory function and is certainly not sufficient for unequivocally demonstrating the existence of stored memories. The above hypothesis of memory-ideas and the known features of brain activity may offer a simpler answer. It is an established fact that all the neurons are more or less active and can thus provide the appropriate inputs necessary to all the circuits alive. (We should perhaps include in the notion of "appropriate" here not only the quality, i.e., the origin, but also the number, a plain sequence, or complex pattern of impulses, which may somewhat limit the activation of the eidetic circuits to some periods of time and to some circumstances.) Again, however, we cannot deny that there may be no need to recall something from registered memory in order to activate eidetic procedures. What is needed is the voluntary focusing of consciousness on those circuits that keep the implemented idea alive, so that mind can consciously read and use it. (The heuristics of the selection is deferred to Part V.)

Rather surprisingly, if we accept a functional role for the constant activity of nerve cells, we have a substitute for stored mass memory, not as it may commonly be conceived in the form of freely re-circulating information or permanently fixed in matter but somehow stored in the form of a continually renewed process of idea ontogenesis. In our hypothesis, the ideas are always existent, always present in our brain, although not necessarily present in consciousness; no one should find this surprising, and in it may lie the "why" and "how" of ideas and memory. Therefore, what we commonly refer to as recalling or retrieving from memory is nothing but the the attention of consciousness addressed to a specific area of inputs or circuits (which is the same). These memories —or better, nonmemories, or simply ideas— are fleeting but stable despite their immateriality; their essential feature lies in a principle of organizing element relationships, ensuring their abstract identity as well as their continual physical renewal. The hypothesis outlined here does not presume to represent an exhaustive framework for the two functions of eidetic system and memory, hitherto considered by neuroscientists as based on altogether different, if in some unknown way connected, processes. However, the hypothesis of a unified framework is certainly more appealing than one positing two separated systems, entailing also separate structures and organizations and a system of functional interconnections, itself with its own structure and organization. The hypothesis advanced here fits the principles of organization of the nervous system discussed in *PTN* but does not represent a logical derivation from it, at least as to the principle of implicit implementation of the ideas.

It is clear that the hypothesis of a unified framework for the eidetic system and for memory does not prescribe the concrete mechanisms (the detailed circuitry) for the functions; this is because at the present stage the mechanisms that one could propose can be represented only by models not yet experimentally testable. As it is, however, the hypothesis may present the neuroscientist fewer insurmountable difficulties than others. Besides the simplicity of unifying ideas and memory under the same neuronal procedures, it does not seem, at first sight at least, to require highly specialized arrangements of neuronal devices for developing the ideas, for transforming them into memory traces, or, particularly, for retrieving the ideas from hitherto unknown traces into neuronal processes such as those which established them. The real question concerning the procedure of eidetic system-memory concerns the way in which consciousness is focused upon the right area of eidetic-memory circuits or to the right inputs.

8.3 Autonomy of Idea Formation

Since it has been assumed that an idea is created (and may be recognized) at the very moment that the elements have interacted in the feedback circuit, the critical factor of the whole procedure, as already suggested, is shifted to the organization of the elements —which may be altogether outside mind's control. In fact, for a large part of the ideas (most of the natural ones) the elements are provided by the outside world, and the formation of ideas in these instances may therefore be almost automatic. In a sense, these ideas are "imposed" on our mind by a procedure that is dependent on the system of relationships among the objects of nature and therefore allows little (if any) role to the participation of mind. The mind finds the elements constituting the objects of nature to be inherently congruent, since natural objects are part of existing reality, and no active participation is needed to recognize their congruence besides the mere acceptance of their matter-of-fact status. Furthermore, the elements of many natural objects in our experience are already ordered in such a way that their intrinsic coherence is also clearly a matter of fact and thus requires only acceptance. Under such conditions, the ideas appear "ready made," as their elements are already ordered and organized in clear-cut systems of relationships without the active participation of mind and in such a way that their "external" order imprints, or imposes, an "internal" order in the eidetic circuits.

I see no reason to reject this view, at least at the present. Natural ideas are obviously of a particular kind, being produced by the constant, direct interaction of the external world with the brain. As such, this category of ideas includes all the memory-ideas that guide the individual in expressing the correct behavior in his own surroundings. In turn, this environmentally dependent behavior is different from the genetically coded one, which concerns primarily only the fundamental species-preserving (and individual-preserving) components of the interaction with the world, rather independently of the concrete surroundings of the individual.

Such an embodiment of many natural ideas that is determined by the outer world and thus virtually automatic opens anew the question of eidetic activity and memory in animal brains. In the preceding section we presented the view of eidetic circuits with a structural substrate consisting of basically simple, although variously combined feedbacks. This type of circuit is generally found in all brain regions and probably represents a fundamental pattern of brain cell organization. It ensures, on one hand, the unity of a function and, on the other, maintains the possibility of extending an influence into other unitary functions, thus to ever larger brain areas, if not to the whole. Animal brains, however, are also organized as complexes made up of the fundamental feedback pattern, probably very similar to the human. Therefore, as we have noted, some raw form of eidetic activity should not be ruled out in animals, in particular the acquisition of something resembling natural ideas, presented ready-made by the environment and requiring little or no active elaboration by part of a mind. If we accept that the development of a quasiautomatic or fully automatic process is possible in animals, we must see animals as also capable of acquiring some knowledge (of themselves, their companions, and the world they live in) in the form of ideas (though not symbolized in language); this would mean that they can exhibit a gamut of interactions with the world (specifically adapted for their own particular, individual surroundings) that is richer than has been foreseen. Still, it does not seem likely that in the absence of logically articulated external and internal languages resembling those of humans (there being no evidence of this), the animal is ever able to develop its raw eidetic processes and exhibit the ability to develop ideas beyond the ready-made stage presented by nature, and even less to construct new ideas. In other words, without the ability for *autonomous* eidetic development, the animal's "eidetic" operation remains at the stage of simple and extremely raw natural ideas.

The lack of the material substrate for language —the only environmentally independent function capable of direct consciousness toward organization of patterns and sequencies of ideas— is the factor that prevents them from autonomously originating those operations of abstraction necessary for developing thought. In other words, while animals also come into possession of some type of ideas, they cannot use them for developing thought; rather, the ideas are used in animal brains as self-organized blocks of "condensed" information as provided by nature for automatically guiding behavior (the whole procedure, moreover, may not require consciousness-awareness).

Man, on the other hand, although possibly acquiring natural ideas in the same manner as animals, makes a different use of them, as shown by the fact that his behavior is largely independent of the environment. This means that for man natural ideas are but one of the innumerable starting points for further eidetic processing —at the most, elements that man uses for constructing other ideas with his language and, by progressively abstracting their contents, for moving away from materiality toward purely cultural regions such as philosophy (this concept is developed more extensively in the "Appendix" to *PTN*). The latter operations cannot be in any way automatic since they are based on generally noncongruent elements, and their specific, intrinsic coherence must be found by "speculating" about their content in relation to the direction of thought, a feature pertaining to the logic-based language of man's rational mind alone.

However, the autonomy typical of human heuristic procedures in developing ideas can also be detected in those natural ideas that are, as we have put it, presented to him ready-made by the external world. Let us return to the example of the container. Obviously, the first time we see a container, the idea of a class of such objects is foreign to us. Therefore, the idea is born only after the elements are introduced, and their relationships are observed, and when we are eventually confronted with a solid block with a hole in it. We use these relationships a posteriori to judge whether an object is a container or a solid with a hole, i.e., we use the ideas to judge about the objects of the world. Nevertheless, it can easily be demonstrated that different individuals use the elements at hand in different ways; everyone follows his own procedure, either by using only part of them or by using them differently, yet reaching the same judgement (the same idea) about the nature of the object. The idea of container may result either from the thinness of the walls or from the diameter of the cavity in relation to overall volume, although by the sense organs all the elements (overall volume, cavity diameter, and thickness of walls) are presented equally to brain. We can shuffle the elements at hand around and still reach in one way or the other the same idea. However, procedures are not all equivalent in terms of logic, nor do all lead to results that are equally easy and immediately graspable, even if from the mathematical point of view the equations expressing the relationships between the elements —and the corresponding analogous implementation in the brain— are fully equivalent.

These observations raise other questions. To explain how different individuals (or the same one at different times) can reach the same result while using the elements differently, it may be presumed either that there is a way of controlling the introduction and use of elements, or that the eidetic circuits themselves are somewhat different, although retaining the ability to reach the same results. The first hypothesis sees natural ideas forming both through automatic procedures and through voluntarily controlled selection of elements, depending on the particular situation. In the case of automatism, the use of different sets of elements for the same idea by different individuals or in different occasions could simply be a matter of chance. However, in both cases the possibility that somewhat different eidetic circuits are involved should be recognized. In fact, a strict similarity of eidetic circuits is impossible in every case due to the practical impossibility of the connections between two neurons and particularly between two such circuits being organized exactly alike. This means that the eidetic circuits can be identified only through their general organization, whereas in the details they may exhibit differences as to the inputs (i.e., the elements introduced) and their relative weight.

We have distinguished between natural and constructed ideas essentially on the differences between their constituent elements. Natural ideas comprise elements with an inherent congruence as a group (a similar general nature) and an intrinsic coherence (the relationships of logical significance binding the specific elements); constructed ideas are made up of noncongruent, i.e., nonhomogeneous elements (their general nature is different), however exhibiting the necessary intrinsic coherence —otherwise their ensemble could not exist as "idea." Let us now enquire whether nonhomogeneous (noncongruent) elements must become in some way congruent for mind to judge about the possible intrinsic coherence making them the constituents of a constructed idea. According to the previous arguments, the intrinsic coherence among elements is established on the basis of two determinant factors that ought to be measurable in some way: a) their belonging to a logical, abstract set of relationships to be represented in an equation and b) their defining, in their ensemble, the structure of a particular desired idea. All this sounds like a postulate of the nature of ideas, because in this way every idea justifies itself through the logical coherence of the elements forming it (even if mind were unable to recognize the fact), leaving understood that if there were no such coherence, there could not be any idea either. But, the prerequisite that the elements must belong to a set bound by relationships to be represented as an equation further restricts them because no equation can possibly be conceived if its elements are not reduced to a congruent set (such as mere numbers or numbers representing physical measurements of events). This means that elements not belonging to the physical world (or more simply, all the elements presenting themselves to mind in a nonimmediate way —those not graspable by the senses or measurable by instruments) must necessarily be "reduced" to a congruent set. This can be done only by depriving the elements of everything contingent while retaining by abstraction those essential features that make their essential relationship recognizable. The reduction to essence represents the necessary step for the set of reciprocal relationships —comprising of these elements a coherent set— to be further determined. Seen in this way, the formation of ideas obeys the general principle that in any system only the relationships (among elements) belonging to the same order, i.e., to the same level of abstraction, can make sense (see also Sect. 4.3.1).

8.4 Voluntary Construction of Ideas

Upon the conclusions of the preceding section we can now reconsider the role of will in eidetic function. We have stated that ideas can be constructed through a voluntary act. This purposeful act involves eidetic activity in two ways: by assembling the right congruent elements that give the desired content to the idea and by constructing the set of elements when the latter are themselves ideas needing to be made congruent. We are dealing here only with constructed ideas since natural ideas from common experience are presented to mind as ready-made abstract objects related to nature's real objects which can in some way be experienced. While one maintains that for every idea connections may be retraced to real objects of nature (including events and natural experiences), constructed ideas do not need to be directly related to experienced natural things since their construction begins from objects already at some level of abstraction. To be able to construct ideas, at least a schematic representation of the goal to be reached must be present in mind *before* the construction, or the assembling, may begin. Obviously, this scheme can derive only from the actual source of the wanted idea: let us call it the "concept" since it still lacks the definite "form" of idea. Put in these terms, will-guided eidetic activity always presumes some preexisting knowledge about objects that mind itself still does not know, but this preknowledge alone does not yet constitute a clear-cut picture of the final product, i.e., a true knowledge complete to the details, such as those already part of mind.

The existence of preknowledge sounds somewhat contrary to common sense; we seek, however, in the last sections of this book to dispell this impression. For the moment we leave this issue aside because there are other arguments to help explain the heuristic procedures for voluntarily constructing ideas. Constructed ideas are generally sought for a well-defined purpose, i.e., to fill the place of what we may call "missing links," as in elliptical reasoning. With this we signify the missing logical connections between individual ideas, sentences, or even theories, either belonging to different regions of knowledge or never before put into relationship to each other in an organized construction. This imposes rather precise constraints on the desired abstract structure since its very purpose requires its integration into an already well-defined framework of thought, bridging the latter's loose ends, to eventually become a part of it. This does not mean that the characteristics of the desired idea are made clear; however, its logical boundaries are defined and constitute an outline, or hull, of something still lacking a precise content but already with more or less definite contours. The preknowledge necessary to construct the idea from an ensemble of available elements must also be uderstood in this sense.

I should not like to convey the notion that in the voluntary process of constructing ideas the quasiautomatic procedures at the basis of a large proportion of natural ideas must be altogether excluded. Quasinatural eidetic motion (see Sects. 4.4.1, 4.4.2) may be said to be automatic in the same way in which the running of a car's engine is automatic when fuel is in it; as in the case of the car's engine, the operating controls and the steering depend on the will and consciousness of the driver. The automatism, in its different degrees, concerns only the way in which the elements and the ideas are presented to mind's consideration. The recognition of (or the search for) the more logical connections, direct affinities, and cause-effect relationships at each step of reasoning are submitted to the ever-present active control, to the judgement and decisions on the part of mind. In this way, mind may be said to select the path and the direction of reasoning and to give shape to new ideas: the procedures can develop logical lines of eidetic activity only if continuously conscious and constantly guided by will.

We now consider the involvement of consciousness-awareness in the procedures of ideas and thought elaboration. It may appear fairly evident that the things may proceed only in a state of consciousness-awareness; however, consciousness, awareness, and the properties of these have not yet been sufficiently examined in the course of our analysis of eidetic procedures. Confronted with the task of giving substance to a missing link in the logic of reasoning, we said that for mind the preknowledge about the content of one as yet unconstructed idea consists of a set of contours defining the boundaries of an empty space in the logical framework of already present ideas. Certainly this hypothesis appears reasonable, but we have also said that further discussion of the problem of preknowledge awaits other arguments (in later sections). These concern specifically awareness, consciousness, conscience, and language, all strictly and reciprocally connected functions —to use a neurological term— but of a still ill-defined nature. Their essence implies that together with the known neurological features we consider the unknown properties and the metaphysical features as well. First, however, we must examine other aspects of eidetic processes.

8.5 Symmetry and Asymmetry of Ideas

All of our preceding arguments have pointed to the conclusion that natural and constructed ideas constitute two separate classes of eidetic objects. Natural ideas are made up of elements that belong to nature and that are inherently congruent, whereas constructed depend strongly upon abstraction and are therefore not based on naturally occurring elements and require a preliminary operation to make them congruent so that the intrinsic coherence may be uncovered. Thus, even if both types obey the general scheme of idea formation, i.e., elements inherently congruent linked together by intrinsic coherence, the final results belong to different categories: those related to physical nature, on the one hand, and those of some degree of abstraction, on the other. As we shall see next, no naturally logical path leads from natural to constructed ideas. Motion through natural ideas does not lead to ideas of a different nature, but remains confined within the natural category. This can only mean that natural and constructed ideas, to say the least, build different phase spaces and may be considered as parts of genuinely different, reference "spaces" (reference frameworks).

The significance of this separation is that two different types of logic are involved in the constitution of the two types of spaces. Natural ideas are objects organized (an organization that we have identified as congruence-coherence of their elements) according to nature's logic. The way in which we appropriate nature's objects presents them to to mind's understanding through the senses, these being the transducers of the object's properties according to causality and time relationships obtaining in nature. Constructed ideas, on the other hand, are objects created by mind's spontaneous operations; their elements are abstractions upon objects of nature and represent the result of autonomous elaboration of nonnatural elements, i.e., abstractions of abstractions. The logic inherent in these operations is no longer related to causality and time (the logical substrate of the structure and dynamics of natural objects) because the structure and dynamics of constructed objects does not know causality and time but only the properties inherent in a single object; each property represents an individuality, a unit element existing as such outside any relationships with the other elements and with time (see Chap. 9 for a discussion of the logic of phase space).

Although we have referred simply to the abstraction "process," we should emphasize that this rests on a property of mind, comprising what is called the intellect, i.e., the capability for abstracting essential significances —the sole basis for understanding the essence of the relationships among nonnatural objects. This capability becomes efficient only with long and painstaking training, as the general capability for reasoning becomes more refined the longer mind is engaged in it. However, while reasoning is developed by logical procedures such as comparisons leading to judgements, abstraction is not reached by means of this or analogous operations because is not concerned with linking ideas or developing lines of thought. Rather, abstraction operates by following a kind of "inward" motion into the "depth" of an idea, —thus quite different from recognizing or constructing relationships among ideas. In some sense, abstraction requires a motion "normal" (or "perpendicular") to every other motion across the phase space of the ideas. In the course of abstraction, an idea is stripped of references to its surrounding objects, of its horizon of other ideas, of other knowledge. What results is the essential content of the idea, however, reached by degrees, each degree being a step nearer to its pure essence. In this procedure there is nothing that resembles a logical operation because logic implies the recognition of necessary links between objects: logic is stating the fundamental principles of these links. Concerning abstraction, the process of recognizing the links among the objects can begin only *after* the idea has reached some advanced step of abstraction, i.e., after it has become a part of another phase space —perhaps also of another reference space system altogether— thus allowing comparison at the same level of abstraction of two initially different objects.

These considerations highlight other essential differences between natural and constructed ideas or classes of objects. We noted above that every idea consists of a set of elements linked by logical relationships. The way in which ideas are embodied in brain matter may perhaps be likened to the embodiment of implicit functions in analogue computers, on the assumption that every idea is representable as an equation in implicit form. We must distinguish the idea in itself from its equationlike representation, as well as from its significance realized in matter, which is different from its embodiment. (We shall see below that an idea constitutes both content and embodiment, i.e., it is simultaneously the abstract object and its reality in physical sense.) Admittedly, the peculiar character of natural ideas lies in they being first-step abstractions (merely to their general attribute of being abstract objects), the mental representation of materiality, after one has included in it all the elements of the same type of materiality and after exclusion of all noncommon elements. However, the term abstraction as we have used it means something quite different since it does not refer to logical operations, i.e., comparisons of elements, judgements about their commonness or noncommonness, but at most to their result. Now, however, a better definition of the term abstraction seems indispensable for avoiding semantic confusion. According to our arguments, natural ideas.are those objects belonging to the realm of nature, and their consistency as a class is assessed on the basis of logical procedures. The procedure leading to formation of the classes of natural objects, however, cannot be referred to as a procedure for abstracting, more properly rather, for "extracting," since it simply extracts the common elements from the mere description of material objects. The procedure requires only the usual comparisons between the sets of properties of the elements which may belong to the class and does not lead into the "depth" of the properties of each object as true abstraction does.

Extraction of the set of inherently congruent, natural elements does not prescribe a particular order of precedence among them, unless one actually wishes to put the idea in equation form. Even then, however, several commutations may be allowed without changing the essential content of the idea. This is the logical consequence of the fact that the intrinsic coherence among the elements may be expressed in different ways, both in terms of reasoning and of symbols; again, characteristic is the previously discussed example of the idea "container." The situation appears somewhat different for constructed ideas. Their constituting elements have been submitted to some degree of true abstraction and reduced to an inherently congruent set. The subsequent logical operation of finding (or recognizing) the intrinsic coherence constituting them as an idea can be based only on the noncommutability of the elements because the coherence among abstractions implies a strict logical ordering, not in a matter-of-factly way and immediately presenting itself to understanding, i.e., an order that cannot allow commutation of terms (of elements), unlike the description of physical objects. Let us take, for instance, the idea of religion. There are several ways express its essence, following one or another line of reasoning, but in each of them we would necessarily consider the idea of God or some equivalent as the common basis and thus primary element of religion; the logical ordering of the elements leading to the outline of the idea of religion cannot be logically reversed, i.e., the elements cannot be commuted.

This suggests a lack of internal symmetry in constructed ideas, preventing the procedure for connecting their constituting elements from being reversed. However, when considering the actual physical embodiment of ideas in the brain circuits, according to the principle of implicit implementation, the noncommutability of the elements is a feature common to all types of ideas since every implicit embodiment is based on a rather precisely defined sequence of operations. Thus, it must be generally excluded that commutability is always allowed at the level of brain circuits. However, this logic of implicit implementation, on the one hand, generally allows a diversity of sequential orders of the embodied elements in the instance of natural ideas without altering the basic significance of the result; on the other hand, it prevents reaching a logical content in the case of constructed ideas. Therefore, symmetry and asymmetry, commutability and noncommutability always refer to the nature of the idea itself, not to its implicit physical embodiment.

Following the standpoint of *PTN*, let us enquire whether commutability and noncommutability of an idea's elements rest on the commutability properties of the corresponding eidetic operators and state ket vectors of the two types of ideas. This in turn raises the problems of the representation and transformation of states at successive times according to the formula of "operator times state function" (with either the moving operator or the moving state solutions discussed in Sects 1.8-1.8.2). In the quantum physics concept of ket vector, the quantitative aspect —the modulus— is not relevant (being reduced to unity), while the measurable property, direction (with a clear-cut qualitative aspect) is the critical factor. To determine the direction one combines all the component vectors of the ket, each with its own coefficient, so that the ket is defined by the ratios of the coefficients. The analogy between ideas and state kets is reinforced by considering, as we have done, that the ideas are unities, characterized only by the qualitative aspects inherent in them —the intrinsic coherence, i.e., the relationships giving a precise meaning to the component elements. Thus, the attention drawn to noncommutability of the idea's elements is relevant, for the conditions of commutability are integral to the operator and state ket hypothesis. Natural ideas, quasinatural motion, commutability of operator and elements are expressed in symmetrical representative equations. Constructed ideas, the result of a voluntary motion, the progressive abstraction of element-ideas, with no commutability allowed between the operator and the elements, affords no symmetry of formulation.

I do not know how much further one can pursue the comparison between conditions of commutability in the procedures of idea formation and conditions of commutability in physical systems at quantum levels. As is stated in *PTN*, the question must be examined thoroughly in advance of further speculation. However, there is still another observation. The essential feature differentiating natural and constructed ideas is the abstraction that dominates in the latter. Abstraction seeks to discover an ever more precisely defined content of a single object rather than the set of relationships among objects; this is thus a motion into the "depth" instead of one "across" the eidetic phase space. The "descending" motion into the elements of an idea is directed by the will and toward an end that is not predetermined, at least certainly not to the extent that the relationships among objects of nature are predetermined by force of their reciprocal quantitative ratios. This may correspond, however, to the basic difference between quantum and classic (nonquantum) conditions, i.e., the relevance assumed in the former by a factor (the Planck constant h) having the dimension of an "action" (the quantum of action). Consequently, the relevance that this can have in either case defines the limit between the condition of necessary noncommutability of the factors entering a quantum state and the commutability of the factors common to the situations of classic mechanics. If one conceives of an eidetic agent having inherent in its nature the meaning of Planck constant, that agent may represent the discriminating factor between the two types of eidetic motion. We can take this agent ever-present in the assembling procedure of constructed ideas, the factor of realization, i.e., of action —and in this case of direction-determination since the object is a vector in a quantumlike system— in other words, the logical factor enforcing asymmetry in the relationships between the operator and the elements of the constructed ideas. With the last sentence we have described the fundamental property of will.

8.6 Abstractions, Will, and Eidetic States

Even if the discussion in the preceding sections can be taken as a general framework for understanding the organization and evolution of neuronal activity as regards the eidetic function, we have by no means reached the end of our problems. All of the mechanisms at neuronal level still require explanation, from the procedures of comparison to those leading to abstraction. Let us consider the latter. One can speculate that the abstraction procedure, as one of the "highest" functions of the human nervous system, is actually much more extensive than it seems at first sight. Besides the messages from neuron to neuron in terms of nerve impulses (as computer instructions in machine language) the brain in all its functions, not only in the eidetic, also uses another language —surely still unknown but nevertheless conveying ideas— of which the impulse-coded instructions constitute only the operative elements. However, this more highly evolved language does not need to be present always to consciousness-awareness, as is required when we are thinking; in other words, it does not need to be "word-symbolic." For instance, the idea of home may be used beneath the conscious level in a host of brain operations connected in one way or the other to the idea of one's own house. The question is whether these abstract representations, at least the simplest among them (for instance, home) can be assembled without the direct intervention of will, by virtue of some automatic force or drive, or whether as low-level abstractions (i.e., extractions) they also are necessarily will guided (see the conclusions of the preceding sections). It is plain that the weight of the question rests on envisaging the likely drives and ways of operating these extractions. A drive to unconscious assembling can be explained by the fact that we would be dealing with classes of ideas of everyday concerns —such as home— created out of sheer necessity, for instance, by mentally portraying first one's closed living space and then upon later experience by extracting the common essence from house, hut, shack, cave, igloo, and whatever else people live in. This hypothesis involves an eidetic motion independent of will, even with ideas at the lowest level of abstraction. The assumption does not appear strikingly at contrast with our previous conclusions if we assume that the same processes set in motion by will are also operated automatically when suitable stimuli are provided; for the structural organization of the brain this does not seem unreasonable at all. (Some unconscious processes may form the basis of the common experience of ideas "never thought of" but appearing at some moment with their almost completely defined content; see, however, the questions concerning conscience, below.)

We can summarize these views as follows: a) By definition, will always is an attentive-conscious process. b) True abstraction motion is a typically voluntary procedure. c) An abstraction motion, at least at the lowest level, i.e., an extraction, can be automatic. d) In order to account for c) one can assume that the same (generally will-guided) eidetic procedure is accessed independently of will by other brain processes. e) The brain uses ideas in all its procedures, also when dealing with noneidetic functions. f) If we assume that some degree of abstraction (extraction) is automatic, then we must ask whether unconscious will, or nonfree will, is possible. g) The difference between a will-guided and an automatic abstraction procedure may lie simply in the depth of abstraction. Will-guided processes can be pushed to the extreme endpoints; automatic abstraction, on the other hand, stops at the level required by the brain processes that produced it, i.e., generally an abstraction at the lowest level, as lower level processes are seen as the unconscious processes. h) If eidetic processes are accessible by general brain operations —at least automatically for the lowest levels of eidetic activity and outside self-conscious control— they may also be possible in animal brain. As noted above, in animal brains the simplest natural ideas are believed to exist only as nonsymbolic abstract structures —somewhat resembling mental pictures— whereas more complex eidetic constructions, such as the true essence abstractions are not appear possible at all for want of a true internal language. i) The extensive use of eidetic processes at some unattentive-unconscious level cannot be restricted necessarily to the lowest level abstractions when the will-operated eidetic procedures are accessed beneath attentive consciousness by psychic processes. Actually this should be very frequent if we assume, as is commonly thought, that a large number of psychic processes, although closely connected with the eidetic function, do not necessarily evolve into a state of directed attention.

The use of eidetic processes and will structures outside the state of clear awareness may surely be contested, but the matter is probably only one of perspective. In fact, it is not very likely that the eidetic processes, having the function of providing the brain with all the representations of the external world and of internal elaborations of ideas, may at the same time deprive the noneidetic functions (somatic and autonomic) of precisely those representations —as are provided by the ideas— that are necessary for their development. This would leave many vital brain functions, at best, with only the insufficient support of primitive, raw, nonsymbolic images, such as those characterizing animal brains. I believe physiologists, psychologists, and psychiatrists would agree on this point. However, this forces us also to accept that an eidetic content is present in virtually all of brain activity, and therefore that, perhaps with the questionable exception of the emotion-dependent ones, all the states of the brain are determined largely by the state of the eidetic function.

While this may seem to overemphasize the role of eidetic activity, we should not forget that language and the eidetic system are the characteristic features of the human brain. Although they appeared last, they also influence the earlier, more primitive, somatic, and autonomic functions —through fusion and the growing together of newer and older anatomical structures and organizational networks, thereby largely fusing their functional properties. If we accept that language and eidetic activity are the characteristic features of man, we should not wonder that this characteristic is also inherent in lower brain activities (those shared with animals). Thus in man the state of eidetic activity is *the* state determinant for the whole brain. This is another way of saying that the change of reference space-time of the whole brain is generally defined by the intellectual activity (the standpoint taken previously, in *PTN*, on the basis of experimental evidence).

8.7 Operators, Transformations, Invariance, and Abstraction

The relationships between eidetic activity and brain states are treated at length in Sects. 1.8-1.8.2, where ideas and operators are related to the general procedure of state transformation according to the formula of "operator times state ket." The property of invariance, typical of the idea, suggests the solution of fixed idea and moving operator, whereas a moving idea would entail something resembling a quasinatural movement, either directed or driven by a built-in, natural, or spontaneously causal agent or left altogether to chance. Obviously, neither proposition is satisfactory. The definition of the time-dependent operator, required by the invariant character of the fixed idea, should be easier than the general operator since at least its time evolution must be linked with the evolution of the ideas in the eidetic process (i.e., with the development of the content). However, we would say, rather, that the operator owes the idea its own time dependence by some reciprocal interactions to an operator-idea-operator continuous feedback, according to the content of the transformed idea. (This concept is discussed in *PTN*; see also below.) At this point we should ask whether it makes sense to speak of transformation of ideas. The answer is in the affirmative if the transformation produces an idea at least logically congruent with the original one (this, however, is still retained as an invariant, integral to mind). We may contest, however, whether one can envisage a transformation eventually producing an altogether different idea, in a way unrelated to the former one, for example, an operator transforming the idea of door into the idea of horse. One can transform the idea hole into the idea house or that of solid body into that of container by adding some elements and finding suitable reciprocal relationships. The gist of this elementary reflection is that by perusing the procedures of thought we can identify the operations performed and can thus hope better to define the operator itself.

In the instance of container, if one had never seen a container before, one might start from "solid body" and add "hole" to it. The idea container arises from setting some limiting ratios between the diameter of the hole and the thickness of the walls, thus transforming the idea "solid with a hole" into "container." Put in this way, for the abstract part of its nature the operator is represented by some functions of the element ideas while for the real, neural part it is identified with the eidetic structure, as a neural operation dependent on the organization of the eidetic circuits. However, the story cannot be considered complete without realizing that besides the ratio, with its symbolic representation and the related numerical solution, there also are qualitative aspects, and that these may also contribute to defining the quantitative characteristics of the operator. Let us consider, for instance, a sponge:

it may contain as much water as a bowl of comparable size, but we never think of it as a container as the bowl, although each of the tiny cavities can be considered a small container by itself and thus seen as a symbolic representation of the bowl. The qualitative aspect is dominant in the case of constructed ideas. Although these may also be defined in such a way that each element is put in some formally logical and roughly quantitative relationship with the others, we would never think of a real, numerical solution of the symbolic representation of constructed ideas as being useful even if feasible. The possibility of a numerical definition is, rather, the characteristic property of most natural ideas, i.e., presenting themselves to mind immediately in the crude symbolic aspect which is prerequisite for a quantitative solution. These considerations also have a bearing on the moving operator of constructed ideas. For these, in contrast to the natural idea operator, knowledge of the abstract part in the form of an equation embodied with the real, neural part in view of a numerical solution, would not be required. In fact, the high level of abstraction of constructed ideas, always at a degree at which any reference to possible natural objects has been lost, actually prevents an exact formulation of the operator's form, except in a few cases when one succeeds in symbolizing precisely the equation-like representations of the original idea through the succeeding steps of abstraction. Still, this equation itself is only an artifice enabling approximate reconstruction of the actually unknown form of the moving operator. The whole procedure is of dubious ustility since it constitutes recognition a posteriori of the form which the operator had *before* interacting with the idea to be transformed, not the form after the interaction. The latter, however, represents the knowledge that is actually sought to "predict" the idea's motion. Moreover, since in the case of constructed ideas the elements are themselves ideas resulting from high-level abstractions, it may be that the operator itself also needs further abstraction, or that one cannot further proceed, in view of the fact that the transformation of the constructed idea might consist in a still higher level of abstraction, a motion for which an appropriate operator cannot yet be imagined (however, see next section).

Let us now return to the formula "transformation = operator times idea ket." Since we have assumed the idea to be invariant and the moving operator to interact in a feedbacklike way with the idea, the invariance of the idea reappears only after the feedback interaction is terminated (since we are dealing with neuronal activities, we can also think of attaining a steady state, or an asymptotic value). At this point, however, the operator itself must also have reached invariance, that is, cease to be a moving operator, meaning that the same operator cannot continue transforming ideas, and that another operator (perhaps, the transformed one, resulting from the feedback interaction with the state) should take charge of further eidetic development. Thus a new idea would entail a new operator which cannot be similar to the preceding one since the idea associated with it would be different, and from this fact would depend the operator's time-dependent character. (This view might contrast with the possible occurrence of contact, or unit, transformations performed by eidetic unit operators, as the latter, producing unitary and uniform steps, probably could maintain a time invariance.)

There are also other considerations. Even if in the brain the real part of the operator —that embedded in the neural process— is necessarily bound to time (in as much as it develops in real time), at the level of the abstract part time itself is not involved in the process of operator times idea. In fact, in the abstract plane the operator, the idea, and their mutual transformations coexist and are immutable as original idea, on the one side, and operator-idea complex (transformed idea) on the other. Therefore, the operator cannot be disjoined from the idea because the latter is a transformed idea only because of its steady coupling (the completed interaction) with the operator, and for this the operator is an invariant. If we accept the interpretation of the final invariance of the operator when the latter forms a complex with the idea, we can distinguish conceptually between the abstract and the real parts of the operator, as is done for the idea. In no way, I think, would it be possible to assume that the real part of the operator —that bound to the neural structure— is an invariant because it is no more invariant than the embodied idea itself and disappearing as soon as the activity of the circuit changes. In fact, when the idea and the operator are embodied in neural circuits activated by different inputs, they acquire a fleeting character, continually subject to an unknown degree of variablity; they are time dependent in as much as they are dependent on the input time course remaining stable perhaps only in a "statistical" sense. The same cannot be said of the abstract part of both idea and operator, time-independent, we may say, by definition.

This leads us to a number of conclusions concerning the duality of both ideas and operators, i.e., the separate existence of an abstract and a real part in them or, if we prefer, their simultaneous existence on two different planes, abstract and real. Only the abstract is truly invariant, as a kind of indispensable insurance against the possibility of variations that may be introduced into the eidetic circuits by other, generally undesired activities. Abstractness is therefore the sole prerequisite for invariance, or is at least sufficient to assure it. On the other hand, one may also see invariance and abstractness as two properties of a single function or as two ways of identifying a single property, or a single essence such as the ideas and the operators are (see also below).

8.8 The Time Operator

According to the conclusions of the preceding section, after reciprocal interaction the idea and the associated operator reach invariance. Thereafter, another operator, one of a different type, generally intervenes to induce a further transformation through the same procedure. This does not exhaust the matter. We know by experience that there are also very slow transformations not obviously amenable to the usual formula of the interaction between operator and idea,
operator we assume another idea or a set of ideas. The change is guided more by the system of complex relationships linking an idea to its surrounding horizon of ideas and appears more like the result of long "brooding" over concepts. For example, my idea of ethics, i.e., that which I consider its essence and its field of influence, has changed in the course of years because over this time the idea has been "tested" with hosts of varying situations as well as with other concepts in psychic, religious, social, and metaphysical areas. None of the changes so produced

is related to a specific operator but to a "maturing" of the idea itself, attributable to acquired experience, more extensive knowledge in various fields, and perhaps, an increasingly penetrating capacity of thought. We shall assume, for the sake of brevity, the existence of a generic time operator, encompassing all the changes occurring in this period, without requiring, at first at least, a more accurate definition.

It is principally or only constructed ideas that are time transformed, but this underscores a feature discriminating between types of constructed ideas. As stated repeatedly, a natural idea may be transformed to another by changing the set of elements. The new idea so obtained does not substitute for the former one, but original and new ideas coexist together with every other idea produced by changing the number, order, or general relationships among elements. Constructed ideas, in as far as they are submitted to transformation by eidetic operators, also are retained unaltered, together with the ideas resulting from further interactions with similar operators. When interacting with a time operator, however, the constructed idea is itself transformed into another idea considered more "true" than the original one and is therefore substituted for it while the former —seen as false, less true, or simply rawer —is eliminated. The conclusion is that the common eidetic operator after interacting with the idea leaves behind two invariant ideas (the original and the transformed), i.e., cannot change the time independence of the ideas. The time operator, on the other hand, generally leaves behind one idea only, the new one, different from the original one. The latter either disappears altogether or is retained as a memory of that which it previously represented —as a "ghost" of its previous self, in as much as it has lost all its power as potential factor of decision in the eidetic process and in the behavior-inducing mechanisms. It is now clear why the time operator interacts only with constructed ideas (and even not with all of them). Natural ideas alone are defined in such a unique way by the naturally occurring relationships among their constituting elements that they cannot suffer time-induced changes —an invariance, then, that is absolutely maintained despite time and internal or external events.

However, this conclusion implies that most constructed ideas may potentially be "killed" and substituted during the process of slow interaction with the time operator. About the nature of the latter, so powerful as to destroy the time invariance of the idea, one may only advance a few speculations. The invariance of ideas is based on the fact that they are abstractions. From this point of view, abstractness assures invariance by excluding every contingent (being the latter variable by definition). One can put this differently by saying that every contingent merges into the immanent abstraction so that nothing can change any more because in the immanent every essential feature of the thing is already included, for it is foreseen in the "ultimate essence" as a possibility. Nevertheless, with both natural and constructed ideas the abstraction procedure is seldom pushed to the limits because every idea is used in reasoning at the level of abstraction appropriate for that reasoning (see Sects. 4.2, 4.3), and this is not always concerned with the ultimate essence of things —-except when dealing with certain metaphysical and religious arguments. As a consequence the slow time operator generally interacts with nonultimate, or rawer, abstractions and eventually succeeds in transforming them into abstractions of higher levels, in other words, in refining the process of definiting essence.

We now confront an extremely interesting question concerning the very nature of abstraction. On the one hand, we have hypothesized that the so-called time operator is not an operator in the true sense but only a name given to the indefinite constellation of factors represented by the wide-ranging relationships of the idea with the outside world probes and with the other ideas and knowledge constituting the horizon of mind accumulated in the course of years. On the other hand, we have concluded that the result of the time operator interaction with the idea is an abstraction of higher level than previously, a refinement of the presumed essence of the idea. These two propositions suggest that what we have previously called the abstraction motion into the depth of an idea can occur only through continuous interaction of the idea with the whole horizon of the mind it belongs to. Accordingly, the search for the ultimate essence of the idea could not consist in an exploration of it in its pure "isolation" but in a continuous confrontation with other essences. This has a metaphysical counterpart in the nature of the essence, as the description of the nature of an object, for it concerns the ensemble of the ultimate properties making a particular *esse* (a particular being) what it actually is. But, the properties are relations, in themselves devoid of significance, i.e., they cannot be considered as individually isolated, and can assume a significance only in relation to their own *esse* or to that of other beings. The process of abstraction (the abstraction motion), representing such a fundamental mind operation that it can lead to uncovering the essence of each object related to its *esse*, an ultimate goal of philosophical research, is a true transformation of the idea. For this, a true operator must exist: not a simple dynamical variable as in quantum mechanics, not a single idea or even a unique principle-idea, but the whole body of knowledge slowly accumulated by the continuous activity of mind.

In this way the time operator hardly differs from the slowly progressing abstraction power. The possibility of a time substitution of ideas is founded on the fact that at every moment our mind can conceive only those relations among the elements that mind itself is able to understand at that moment, and no others. Therefore, since only by slow training in eidetic activity can the mind's ability to understand relationships become enhanced, the power of its abstraction procedure can only slowly be refined to the point at which the ultimate meaning (nearest to essence) of things is present at its clearest. (The change that the idea of physical space has undergone from classical mechanics theories to general relativity represents one of the more striking examples.) From these arguments we should accept, in contrast to a previous statement, that in the realm of nature (not of the plain natural ideas but of the constructed ideas concerning the essence of nature) it is possible for time operator-dependent substitutions of ideas to take place since training in scientific thought and slowly acquired knowledge are the only natural ways for increasing mind's horizon of objects.

To summarize, natural and constructed ideas differ as to their mode of abstraction motion. The mind-independent relationships among the elements of natural ideas can be represented as formulas of clear symbolism, as physical laws. The abstractness of the physical law assures invariance because, representing only essential relations among material elements, it does not depend upon physical measurements of the latter and does not make any reference to particular situations or time. On the other hand, the abstraction at the basis of constructed ideas cannot always, immediately, or easily be constrained into a symbolic representation, such as a mathematical formulation of the relationships among the constituting elements, although this surely could be done. These ideas generally result from progressively higher levels of abstraction attained through a process of going deeper into the essence of an individual idea, a procedure not easily representable in symbolic terms nor conceivable simply as reordering a system of clear-cut relationships among the elements of that individual idea. More specifically, however, the higher levels of abstraction are reached through the interaction of the idea with the objects forming the wide horizon of mind's knowledge acquired in the course of years, i.e., with that which we have called the time operator. This does not introduce a time dependence in the place of the invariance of the idea but to express the same concept substitutes for the present another idea at a more advanced level of abstraction. The invariance of these abstractions is founded upon the "necessary" character of these eidetic objects which strive to render more completely the content of a still ill-defined concept. These abstractions are a sign of the state —or level— reached by mind's abstracting power, in other words, of mind's ability to understand the complex relationships among the elements comprised in that concept —we might also say, the sign of the limit reached by the capacities of mind and therefore an "absolute", i.e., an invariant.

These arguments portray the time operator as having the general character of maturation, revealing an enhanced ability of mind to understand not simply the surrounding nature but principally its own constituting elements. The ideas themselves mature similarly as an individual does, not leaving behind his former self as an invariant image but substituting for it another and different self, leaving nothing behind but a faint ghost, an echo, of the past essence. These are *evolving* ideas, and the time operator is the factor producing the aging or evolution. In this restricted sense only may these ideas be said to be time dependent, although generally the dependence on time develops over such a long period that it is justified to regard them as invariant. The rationale for this apparent invariance is that the procedure of substitution of one new idea for the older one is the unceasing mental maturation (in whatever form defined), but this in turn consists of an unceasing, if slow, change in the whole brain state, or reference system. According to our theory of brain state changes, mind because of the time operator is no longer able to produce idea A but only idea B on a level of abstraction higher than A. In doing so, the time operator deprives mind of the possibility of a fixed time reference for detecting its own change, if not the existence of those faint ghosts or echoes of a lost world of ideas. (In this image, the time operator, though a slowly acting state transformation operator, is still connected to the general theory of the eidetic operator discussed in Sects. 1.8-1.8.2.)

The view that the eidetic system —actually, the ideas themselves— represents the principal source of transformation of the brain's space time, implies that the ideas are tightly anchored in the cerebral structure (and its activity) and maintained in a set of relationships resembling those described in the theory of general relativity. The same type of relationships can also be taken as the basis for a more metaphysical interpretation of the relations between state of the the ideas and the brain's structure. Pursuing the relativistic reasoning, the ideas —abstract and invariant structures— are outside the influence of time as physical clock time, and are subject only to the cerebral "personal" time, connected to the proper time of the system. However, according to our theory (based, after all, upon some convincing experimental evidence), the proper time is determined by the ideas themselves, e.i., by the eidetic processes in general. Thus, in some sense, the ideas are also outside of time as they themselves are the origin of brain time. In terms of relativity theory as a general framework for brain activity, including the eidetic, the relationships between time and the ideas in brain space time are seen as the relationships between time and matter-energy in physical space time. In the latter, energy and matter, an invariant, define the temporal and spatial parameters; in the brain the ideas, which are also invariant in constituting their own reference framework, likewise define the temporal and spatial parameters of that framework.

Part V

Origins of the Eidetic Function: Concepts and Language

Part V

Origins of the Eidetic Function: Concepts and Language

9 Consciousness, Awareness, and Conscience

Part V deals with the origins of eidetic function. These origins are sought not in physical structures but in complex abstract objects and their logical interrelationships. To be experienced and used, the abstract structures must possess a physical substrate and thus participate in the physical life of the mind-brain; this principle was discussed at the beginning of this monograph. The origin, however, must be abstract if one is to gain knowledge of new relationships among abstract objects, that is, to form new thoughts, independent of a primarily, physically based mind experience and due to the high-speed appearance of fleeting abstract objects of thought. These are the two fundamental properties that cannot be accounted for by any theory of the eidetic function based exclusively on a physical substrate organization and its related functional properties. Obviously, the physical substrate cannot be dispensed with in the process of thinking, i.e., throughout the thought's mental development over time; the introduction of the time of development is tantamount to the involvement of the time-dependent activity of the structures of the physical brain. The thesis in Part V is that the physically conceived mind is insufficient to explain all the properties of the eidetic function, and that the truly primary origin of thought must lie in nonphysical, essentially abstract objects. This view sees the physical mind as but a faint reflection of its abstract counterpart.

In the discussion below we define the structure and operations of the abstract side of mind on the baisis of an analysis of the relationships among awareness, consciousness, and what we have referred to as phase space of the ideas; the latter we shall redefine as conscience (without any ethical or religious connotation). We shall attempt to demonstrate the necessity of the "real" existence of an abstract phase space (since there can be no physical demonstration), that is clearly distinct from a physical phase space which can be identified with memory (in whatever form the latter might be envisaged).

9.1 Ideas, Consciousness, and Awareness

We have, up to this point, left unanswered certain questions concerning the independence of ideas from time. Such independence from time may seem problematic since to be used in thought an idea must be implemented in a neuronal circuit and is therefore subject to the temporal development of the latter's activity.

One cannot simply liken the implementation of an idea in brain circuits to the writing of a formula on a sheet of paper. In the latter the idea expressed through the formula already exists in mind, i.e., in brain circuits, whereas in the former the idea does not exist until it is implemented in matter. In both cases, however, the independence from time (in the sense of invariance) rests on the fact that in the idea the parameter of time —as a reference— does not enter as a determinant in the formulation. Even if in the neural circuit the idea is formed in a sequential order of operations developed at a velocity much lower than in electronic circuits, time is instrumental only in the completion of the material operations and cannot influence the relationships among the elements constituting the idea. Moreover, the idea does not exist as such until the implementation is completed and the circuit is closed, i.e., only when the operations are completed. This applies to ideas stored in memory —if such a storage indeed exists— because their recalling and becoming concrete in mind still require the time of the corresponding neuronal operations for performing them; once the process is completed, however, the ideas present in mind are the exact representation of the stored "template" and are in every way independent of time.

The temporal independence of the idea may also be seen as in a more profound sense. The impossibility of comparing the implementation of an idea in brain circuits to the writing of a formula on paper draws our attention to a major gap in our discussion of the formation of ideas. We feel that, as in the case of writing a formula, the idea should be somehow, present, if not in full at least in outline, *before* its implementation; there must first somewhere be the concept to which the idea simply provides a form. The rationale for this is that without some knowledge of what the idea should look like, the idea cannot possibly be implemented in the correct way —unless one sees the implementation as the primary act, the "origin" of the idea. However, this also would imply that since there is no previous knowledge, the procedure always consists in trial and error —the same situation as with a monkey typing letters at random, waiting for a poem to emerge eventually. We are thus confronted with a problematic task here.

That ideas are independent of time does not mean that they are necessarily outside of time. However, the rapidity of most eidetic operations may appear to suggest the simultaneous presence in mind of many if not all abstract structures. If this were truly the case, the arguments at the end of the preceding section would be further strengthened. In fact, ideas are abstract structures implemented in brain matter and as such, according to relativistic principles, are not influenced by physical clock time but are subject only to the brain's "personal" time (a function of the system's own time, i.e., of the *Eigenzeit*). However, there is some evidence that this time of the system originates in eidetic processes themselves (see the discussion in *PTN* on this and the preceding section); this would suggest that the ideas are outside of time in the sense that the eidetic processes are the origin of brain time. This posits therefore a parallelism between physical world space time and brain space time, both of which having matter-energy as the determinant of spatial characteristics. Strictly speaking, we cannot refer to ideas being outside of time if it cannot be demonstrated that they develop with a velocity greater than the limit velocity (that of light in the vacuum) in physical space time. However, eidetic operations, involving objects having an abstract structure, could themselves well represent the determinant of the limit velocity in our brain and therefore the determinant of the unit of measurement for personal time (as in physical world the velocity of light). Unfortunately, we have no way at the moment to test this hypothesis, although it is certainly in line with the preceding arguments and with the theory in *PTN*. Thus, in principle, simultaneity in brain space time would have no exact physical meaning —as it has no exact meaning in physical space time— since all our means for detecting it are bound to much slower velocities, even if the neuronal limit velocity is considerably lower than that of light and therefore more easily accessible to experimental control.

The latter observation suggests that that which is impossible in physical space time principally because of the enormous distances involved is also impossible in brain space time, due to the slowness of the communications, despite the infinitely smaller distances within the neuronal circuits; there would always be an inadequacy of velocity in relation to space. A neurophysiologist would probably see no problem in all this initially; however, multiple probes placed at strategic points within the nervous system would allow detection of the simultaneous presence of ideas —better, the simultaneous embodiment of ideas— in different, separate circuits, as Maxwell's demon would be able to see the simultaneity of events at cosmic distances. This procedure would nevertheless offer only a sham solution to our problem because the mere presence of a multiplicity of electrodes still would not endow the brain with the capacity of gaining awareness of the fact and therefore of using that simultaneous knowledge to its purposes. Even if we could find a way of by-passing the slow conducting connections between neurons by means of high-speed (electronic) conducting devices, faithfully reproducing neuronal messages, we would still never be able to transmit the desired information through the correct constellation of synaptic activities on cell membranes with sufficiently high speed as required for detecting simultaneity due to the long time constants involved in the excitation of these structures. In each case, however, the procedure would create an overly artificial, unnatural situation, and one would not be justified in basing any conclusions of physiological value on it.

While these arguments against the hypothesis of simultaneity cannot be dismissed, at least on purely theoretical grounds, we must also recall certain properties of the nervous system which tend to change the perspective of the problem. One of these is the fact that each neural function takes its own time to be developed, in *PTN* we see this as representing the temporal characteristics of a neural function, completely unknown at the present but certainly representing some finite if small, measurable period of time. This is in general state dependent, for it is influenced differentially depending on the actual state of the brain. Important here is that these characteristic times are within the physiological range of a few tens or hundreds of milliseconds, and that simultaneity must also not be seen as absolute but within these ranges. In fact, every judgement on simultaneity depends on the characteristic time constant of a hypothetical subsystem gaining awareness of the events at different places within the brain. Therefore one cannot exclude the concrete possibility that *in practice* the slowness inherent in the nature of the connections may still allow us to judge as simultaneous the presence of ideas at different places in the eidetic structures or in the whole brain, provided the incoming information concerning the ideas is timed to fall within the characteristic time of the function. On the other hand, the same possibility of constructing new ideas rests on the simultaneous existence of ideas and elements at different places within the brain; we are aware of all of them, and we use them as elements in building the new idea. The implicit embodiment of the constructed idea, based upon a closed loop of complex composition as to number of element-ideas, permits the recognition (and practical use) of a true simultaneity of events within the nervous system.

9.1.1 Consciousness: A State and an Idea

We should point out that in considering the possibility of the simultaneous presence in mind of different elements and ideas our mind must have awareness of them to be able to construct other ideas. With this observation we confront the problem of consciousness. Scientists from various branches of neuroscience have defined consciousness in different ways. A unitary definition is needed here which integrates the peculiarities from all areas of neuroscience; the metaphysical aspects must also be regarded as inherent in the overall concept of a state, function, and set of properties.

To start with, let us consider consciousness simply mind's general attentive state and see mind's functional presence only in the flux of events inside the brain, a simple and unconcerned presence that disregards the attention to specific objects and mind's engagement in the flux of particular events. To take an example from Husserl's phenomenological analysis, I can be conscious of seeing while the object of my sight does not partake of my being conscious of sight, although I am conscious that my consciousness of seeing may also comprise all those objects that could possibly be seen. In the same way I can also be simultaneously conscious of hearing, thinking, etc. because these functions are simultaneously present and together form the background of my mind's being present and alive. Seen in this way, consciousness, on the one hand, has a wide embrace since it comprises everything that may affect the senses as well as everything constituting mind's knowledge. On the other hand, consciousness has a very limited depth, for this excludes that the presence of all the functions —seeing, hearing, etc.— also imply the presence in mind of the specific sensations associated with them as well as of the abstract objects —ideas and concepts— pertaining specifically to the region of each of these functions, i.e., sensing and thinking are excluded.

On the basis of these considerations we can define more precisely the other aspects and properties of consciousness. Since consciousness presents itself as a mind state, one can ask, first, whether it is also an abstract structure, a product of the same mind. This is not an idle or rhetorical question. The hypothesis appears supported by the fact that the abstractness of consciousness is ensured by its being independent of that which is contingently seen, heard, or thought, i.e., independent of the neural processes initiated by sensory experiences, up to those of idea formation and reasoning. Because it is independent of contingent objects that may concern the mind, consciousness is left with nothing to be concerned with but its own presence, and it may therefore also be considered independent of time, the realm of the contingent. Seen in this way, consciousness is a truly abstract structure because in its simplest sense it represents the idea of being simply conscious; this is its essence. There is a peculiarity in this idea, however, that makes it unique. Consciousness represents something more concrete than a simple idea. The fact that I can be simply conscious —by emptying my consciousness of any content, i.e., by eliminating the contingent and its related time— reveals that consciousness has a very particular nature, and that it must always be simultaneously idea (the idea of being conscious) and the concrete existence, or realization, of its content (its real nature) in the brain structure, without which one could never be simply conscious. In other words, one could not feel conscious if the idea consciousness were not embodied as idea in brain matter and rendered real in its content, i.e., unless the state of simple consciousness is also realized; this means that the idea "I am conscious" is the same as my actually being conscious and cannot in any way be separated from it.

Continuing this line of reasoning, we can observe that every other idea exists for me —that is, I can think the idea— only if the idea of consciousness has been embodied and realized in the brain circuits, and with it I have become automatically conscious, i.e., only if consciousness has become a state reaching awareness. I may think of my being conscious of hearing, for instance, without actually hearing anything, as well as I may think of a dog independently of any materially existing dog; in these instances the idea and the simultaneous physical existence of its content are two separate entities. For simple consciousness this is not possible: the idea of being conscious and the state of consciousness are the same thing, and we cannot say which comes first in the material substrate of mind (if it is possible at all ever to pose such a question), because at the root of every possible neural mechanism evoking consciousness there may well be hidden the idea of consciousness. (This argument was used above in advancing the belief that the brain uses ideas at the subconscious level while performing different functions in the somatic and autonomic domains.)

However, other arguments argue strongly against the view that consciousness can be considered an abstract structure in the sense described here. One might argue that consciousness is not actually independent of time on account of the fact that merely its being embodied in brain circuits makes it participate in their fortune, i.e., subject to very rapidly disappearing and reappearing at short intervals —thus, if not truly a time-dependent entity, at any rate one somehow related to time-dependent situations. In other words, consciousness is always linked to the content of the flux of events. Again, as a state, consciousness is commonly conceived as dependent on the mechanisms controlling the general state of the brain and therefore not necessarily under the control of mind. Furthermore, consciousness as the state of being conscious is an act —and therefore no longer an abstract function— implying a "here and now" that is awkwardly disjoined from an attention to something and in the present. In fact, the present is the only time at which attention is possible; for instance, I am conscious as well as being conscious of sight, of hearing, etc. only in a continually renewed here and now.

As in the above examples, consciousness, although independent of the contingent, is always linked to an attribute indicating the region of consciousness, such as consciousness of sight, hearing, etc. We could hardly find a meaning in consciousness either as state or as idea of being conscious, if not accompanied by at least the more elementary of its possible attributes; accordingly, there is no actual consciousness —as act of being conscious— if not accompanied by at least its specification of being conscious, the more reduced meaning of simple consciousness. On second thought, however, consciousness, either as state of consciousness or as idea, cannot be separated from its special connection to time. Consciousness as state introduces into the nature of the idea its own feature of the "now" and constantly renewed "present" where events flow unceasingly As we have seen, however, consciousness may also constitute a category by itself, not only as both idea and its realization but also regarding its relationship to time, for in the embodiment of the idea and in its realization the presumably time-dependent neural processes may be precisely those which define the unit for measuring the time of brain processes. We should, therefore, talk of a reversed dependence i.e., that of time (the brain's "personal" time, or the system's *Eigenzeit*) on this unique, special type of neural event. Here, consciousness represents the necessary link between the abstract and material realms because it unifies the properties of abstractness and materiality. In this view, consciousness is like the door through which every abstract structure must pass to become real in mind; in developing thought the ideas are "realized" in the brain eidetic circuits for the very reason that we have "realized" consciousness.

9.1.2 The Time Constants of Consciousness and Awareness

We have observed that it is possible simultaneously to have simple consciousness of sight, hearing, thinking, etc. Consciousness, however, is not naturally so uncommitted (i.e., not simple); it selects or addresses itself (or is addressed) to a region and its specific content. Since the abstract content (the objects) of any region as a whole may be simultaneously present to consciousness, although not yet individually explored, again we are confronted with an immediacy or instantaneousness of becoming real that remains unaccounted for by the velocity of the neural events in the brain circuits. Therefore, consciousness could imply "infinite" velocity not only when uncommitted but also when using an "address" or "lens" to access a particular region. Furthermore, consciousness, when no longer using the lens in that region, either vanishes as consciousness or uses the lens in another region, with the same simultaneity of transition. In this perspective, consciousness always temporally precedes awareness (the state of the exploring mind), and the latter always presupposes logically, i.e, necessarily, the former.

It is proposed in *PTN* that the reference velocity c_{CNS} (the velocity of some neural parameter in principle analogous to velocity of light, c) could represent the relativistic term of comparison for the velocities of all the other neural events. No definitive hypothesis was put forward, but not having to account for the abstract structures and their motion a matter-related velocity acceptable as a reference was possibly envisaged in the highest rate of appearance of the nerve impulses at the synapses. Now that we are considering the velocity of eidetic structures, such as those involved in assembling the idea of consciousness from available element-ideas, the c_{CNS} velocity can be represented by the velocity of those processes which seem to be reaching the limit of simultaneity of presence within the central nervous system. This hypothesis is at a variance with that put forth in *PTN*, but perhaps the two can ultimately be unified. The general mechanism of assembling constructed ideas from element-ideas simultaneously present in the maze of the eidetic circuits cannot yet be described in such simple terms as with the impulses impinging on individual synapses. Rather, in our theory, the neural messages associated with element-ideas must be viewed not at the level of the individual neurons but that of the circuits where new ideas are assembled. Therefore, the time constant dependent on the velocity c_{CNS} may no longer be that considered in *PTN* but that of the eidetic circuit at the very time the new idea is born. Thus, it is possible that the two c_{CNS} values coincide if one takes into account the velocity of implementation in the neural circuits, encountering its limits in the time constants of the neurons's membranes.

I suspect that we are confronted here with two distinct situations involving different velocities. In the one is consciousness in the sense of general, simple consciousness (of sight, hearing, thinking, etc., all simultaneously present). In this case, since consciousness is not concerned with specific individual contents, sight, hearing, and thinking exist simultaneously in consciousness. (It should be studied whether consciousness, although the unique and unitary function-state of a specific self, can remain in that way undivided between the two modalities. Resuming the line of thought of *PTN*, we could hypothesize that the rationale of the situation consists in consciousness being at the same time in different states, such as subatomic particles, in analogy to quantum mechanics principles.) In the other situation, consciousness focuses the lens on one region, not simply regarding the simultaneity of presence in it of the various objects but now perceiving the presence of the individual objects one by one —e.g., sight— by exploring that region. In this instance, the state can no longer be that of simple consciousness but is more precisely called awareness, to distinguish the difference between the two situations. This means that when different objects exist simultaneously, simple consciousness regards only the simultaneous existence of a variety of objects but of none individually —a very high velocity motion. On the other hand, when one examines individual objects, we call the situation awareness, which alone allows gaining knowledge of them, however operating only "sequentially, i.e., in real time, implying that although as rapid as we may want, it excludes the simultaneous consciousness of the individual objects. The origin of this second situation is marked by the moment of transition from an almost infinite to a finite and very low velocity (see below), i.e., the moment when consciousness, becoming awareness, uses its lens to look at things, when the objects are taken one after the other, and for each the state of awareness is undivided. Both the slowing of the motion and the undivided state of awareness can be attributed to the interaction of the eidetic objects with the eidetic operator (see Sects. 1.8-1.8.2, 8.7, 8.8). As in quantum mechanics, when the particle interacts with the measuring instrument, one can no longer say of it that it is in different states simultaneously but only in that one allowed by the experimental conditions.

9.1.3 Investigating the Mechanisms of Consciousness'

The basis of the reasoning above lies in the fact that interaction with the eidetic operator brings about a clear-cut change of state, such as the transition from simple consciousness to awareness. In the new reference framework (according to the theory in *PTN*, each state represents a reference framework for every event occurring in that state) local geometries are substituted for previous —also local—geometries, i.e,. new units of measurement of space and time are substituted for the former ones. If we still conceive of consciousness as a state (with certain peculiar characteristics) of brain matter in which neural events occur in the form of nerve impulses traveling along fibers and of transmembrane potential oscillations within the range of electroencephalographic frequencies, the physical brain space parameters of consciousness and of awareness are likely to reveal invariance in the two frameworks, and only the time parameter may change. Consequently, if in consciousness the eidetic events seem to develop at such a high speed that we are led to talk of simultaneity, whereas in awareness the velocities are much lower and within the easily measurable range of neuronal conduction across cell membranes, the conclusion can only be that the personal time of the two reference frameworks must be different, with the personal clock of consciousness running at a much slower rate than that of awareness.

This should not be taken to imply that the speed of nerve impulses is different in the two frameworks (see also *PTN*). In fact, we cannot measure different velocities of nerve impulses in the two states; only a scaling up or down of EEG wave frequencies may be hypothesized. We should remember the fact that the two states, although states of the whole brain (since eidetic activity involves practically all of it), are actually psychic states, i.e., states of the *Ich* (of the *Ego*) and as such are not necessarily reflected in the physicochemical, basic events of the neuronal matter. We should expect, instead, the difference between the frameworks to be manifested only at the level of the individual *Ich* and not directly observable (nor thus measurable) by an external observer, even if equipped with sophisticated physicochemical measuring devices. However, the question cannot be dismissed so simply.

We can advance two hypotheses. a) Since they are specific manifestations of the life of the *Ich*, the two states are themselves the reference space of the *Ich*. The latter therefore cannot directly gain knowledge of the difference between the two time references but only that mediated by observing an external clock (that is, a clock at rest in another reference framework). In its own space, i.e., in its own state, the *Ich* can only be conscious that the ideas sometimes exist "all at once;" they appear in the mind with immediacy, and even in "blocks," as though skipping the intermediate steps in idea formation; at other times, eidetic operation proceeds slowly, exploring a particular region stepwise and forming or selecting the ideas individually one by one. The two situations correspond to the two relativistic properties of time in the two states, consciousness and awareness, properties that can be detected and investigated only with reference to an external clock, not with the other physicochemical tests of the physiological laboratory. b) Alternately, one can posit that the difference in time reference at the basis of the physiologically different properties characterizing the two states is itself determined by some physicohemical process or phenomenon that may justify the peculiar property of consciousness, namely, the characteristic immediacy of its manifestations. If this were the case, the relativistic time difference would be the direct "spy" of that process or phenomenon, and suitable physicochemical tests could be found to detect and explain the operations of the *Ich* in purely chemicophysical terms.

The first of the hypotheses affords only psychological approaches; the latter allows only those which take into account —or, which create— situations of the two states characterized by an accurately controlled presence of relativistically behaving parameters. Therefore the investigation must be centered about the set of parameters constituting the desired situations and on the assessment of the correspondingly induced relativistic time differences. At first glance, the second hypothesis offers a variety of approaches corresponding to different working subhypotheses. One may assume, for instance, mere field effects arising from any of the many physical and chemical events occurring in the transition from nonconsciousness to simple consciousness (hitherto poorly investigated). Consideration of the hypotheses pivots particularly around a) an instrumental role of brain electrical activity and b) signs of the presence of consciousness possibly revealed in some specific frequency band of brain electric potential waves. The latter sounds reasonable on account of the well-known EEG correlates in the transitions between states. (An interesting speculation that would unify the two alternatives has been advanced by D.Rakovic and his collaborators in recent years. This is noted here because it is theoretically sound and has an appealing simplicity. The main point is a clever explanation of the unexpected role that would be played by the electromagnetic component of brain EEG waves propagating through a low-dielectric structured medium. Without going into detail, this model accounts for a relativistic proper time —*Eigenzeit*—dilatation, which would explain the sensation of suddenness in the manifestation of the state of consciousness, together with other properties in certain particular spontaneous and induced psychological states.) Until this situation has been better analyzed, we should avoid further speculations, pointing out only, once again, that the idea of consciousness and its realization seems to belong to a special category of eidetic objects, possibly with an associated, unsuspectedly high velocity.

In conclusion, consciousness is an abstraction, an idea with its property of invariance (time independence), but it is the state of consciousness as well, since in this case the embodiment of the idea is the same as the realization of the state itself, i.e., as its material "reality." Other ideas and their embodiment belong to well-separate categories; they are abstract structures —although also embodied materially in brain circuits— without their content actually becoming material. (As has been said, embodying the idea of consciousness is our "being conscious" whereas the embodiment of the idea of seeing does not make us see.) However, the peculiarity of consciousness is not limited to its being a category apart; it also plays a role, for at the same time it is a function of the central nervous system. We feel strongly driven to consider consciousness as a transition or communication structure, bridging two spaces (two real worlds): that of the structures without time, the phase space of the ideas, and the space-time of the physical world. In fact, consciousness in its most general sense (thus not in the sense of awareness focused upon a particular field), can live only in the present because it cannot be consciousness even of the flow of time; therefore it can be said to be a timeless function since without a limit to the infinite smallness of the present, the present is "no time." For the same reason, consciousness lives only for the timeless duration of the present and must be continuously renewed in order to be always the same. Its function, or its instrumental role, consists in the fact that only through the state of consciousness is the world of ideas, of abstract structures, transferred from the timeless region (the ideas' phase space) into the physical space-time of the brain. The slowing down of the rate of motion from consciousness to awareness is necessary to fit the rate of ideas' embodiment through the neuronal circuits of brain matter, to label them, read them, and organize them into the language of thought. Reversing the path, the elements embodied in brain circuits may be assembled in the state of awareness and their coherence assessed; with them ideas are formed, and through the infinitely small window of the present —the timeless consciousness— they can be subsumed under the abstract structures of the timeless region, thus becoming part of the ideas' phase space.

9.2 Implicit Implementation and Consciousness

The high-speed appearance of ideas in the state of consciousness is a peculiarity of this state, making it an altogether different phenomenon from other, noneidetic neural events developing with much slower velocity. This does not exclude some aspect of eidetic motion contrasting with the high speed of eidetic operations in general, for instance, in the evolution and construction of ideas. These are gradual processes lacking the flashingly sudden appearance of ideas in mind while thinking. In evolution and construction there is a temporal sequence of events, thus slow development, of sequences of accepting and discarding elements, of trial and error procedures, of analysis of concepts and synthesis of judgements, until the idea have reaches completion, i.e., the set of procedures that is typical of awareness. These manifold aspects of consciousness must now be considered in terms of our previous conclusions as to memorization and will. The field idea represents a finite memory block (in some form) containing ideas differentiated from the idea core, all simultaneously present. Consciousness seems to meet no difficulty in instantaneously focusing the lens on a particular idea within a field; only when the purpose is subsequently to examine the idea does consciousness transformed into awareness lose its flashing immediacy. The example of the field idea is adduced as a paradigm; in fact, nothing would prevent the same argument being used generally in consciousness focusing upon any of the different regions and types of ideas. We now accept explicitly that in any definite region the respective ideas are present all at a time, and even if some of them represent a development of others, it is possible to focus upon them with immediacy. The basic concept is that of phase space already as discussed in preceding sections, and this will draw our attention for the rest of this monograph.

Our arguments now begin to enter the area of metaphysics; as we have noted repeatedly, consciousness is both an abstract object and its realization, i.e., a unique, indivisible, simultaneously abstract and material object, which is also a state of the brain. We have described consciousness as being connected simultaneously to the space of abstract objects (the phase space) and to brain's time and matter space. In this way, consciousness senses, if one may say so, the simultaneous presence of the abstract objects of its type and lets them "flow" through its own timeless present into the matter and time of brain to be embodied there. Surely, this is "applied metaphysics" in its etymological sense, but I would like to add that if the metaphysical properties of the brain are considered a logical necessity, then they become as real as the physical ones. I am convinced that we can never otherwise explain the nature of ideas, mind, thought, knowledge, and even less conscience (as we shall see below). We may refuse to recognize that some metaphysical properties are integral to our brain function, thus more or less explicitly assuming that this metaphysics is an artificial construction, a mere physical product of brain cell activity. The question may be answered (if ever possible) only by searching —as we shall do below— for the link between the metaphysical and the physical aspects of brain life and for at least a hint of their ways of operation. Clearly, the main question involved concerns the assumption of a "real" (in act) phase space of the ideas.

The concept of phase space is borrowed from quantum mechanics and represents the 2n-dimensional ordered space consisting of the Lagrangian coordinates of position and momentum of each systems' element. In our case, the first obvious material "realization" of phase space consists in the ensemble of the Lagrangian coordinates of the neurons constituting the implicitly implementing circuits. Although the neural circuits implementing abstract objects are obviously recognized as being formally not identical to the circuits implementing implicit functions in analogue computers, they have been assumed to have in common the principle of implicit implementation. According to the latter, the required elements (in our case, of an idea) must be present at the input of nerve cells suitably linked in a feedback configuration, thus implementing the idea. Important here is that both the elements and the idea are come into existence together, at the very moment that the feedback circuit is closed, and keeping the circuits closed is what keeps the phase space alive. The scheme of the eidetic mechanism realized in matter, with many inputs turning the process on and off, makes it suitable for being continuously active, probably even outside consciousness. Certainly, the ideas could always be there even when awareness is not using them but still being used in several neural processes not requiring awareness. It is logical to assume that an eidetic operator enters the picture precisely at this point. In fact, if we accept the ideas as being implemented in brain matter and therefore invariant in the structure (a kind of physically based memory, as noted above) besides being invariant as abstract objects, the eidetic operator would be called for only to focus the "lens," in other words, to direct awareness onto this or that circuit —i.e., this or that idea— one after the other, in a temporally and logically ordered sequence.

The transformation brought about by this operator concerns only one aspect of the eidetic function, namely the transformation of consciousness into awareness. If one also accepts that the physically embodied ideas, i.e., the sets of all the neurons involved in representing the ideas, in their ensemble constitute a realized phase space, then the latter —since it is a phase space, and notwithstanding its embodiment in brain matter— must be in some way independent of time so that the operator actually activates the transition from what we have called the timeless region of eidetic operation to the eidetic operation of brain space time, from the flashing shifting of consciousness to the infinitely slower velocities of neuronal motions. More than other transformations of brain state, that operated by this eidetic operator recalls the transformations of the overall brain reference framework, since what is aroused is a specific eidetic "attention" allowing the perusal of precisely identified objects, in place of a general consciousness of simultaneity of "presences." We cannot say that in the transformation the metaphysical connotations are eliminated, because the phase space as such still remains an abstraction. How the operator is guided, and how will enters the eidetic transformations, are questions to be discussed below. First, however, we must introduce other concepts and arguments more specifically metaphysical and place them in a more comprehensive context together with others which strictly relate to the physical world.

9.3 Immanent Ideas and Logic

Let us now consider the picture of "immanent", "stand-by" ideas, continuously present and embodied in the brain structure organization, waiting only to be utilized. How utilization may occur is first of all a question of the relationships between ideas and consciousness. The ideas may be immanent in two ways: a) implemented in a feedback circuit with the feedback path closed, therefore immanent and actual, although awareness may still not be concerned with them; b) implemented in a feedback circuit open at some point, therefore immanent only in potency, but not currently. Regarding the utilization of ideas in these two conditions we can also put forward two possibilities. On the one hand, due to the continuous activity of neurons, the feedback paths are practically always closed (activated), so that at virtually any time consciousness may focus its lens with suddenness and become awareness, which "reads" the idea and connects it to others. On the other hand, due to the fact that the feedback loops may be very complex, particularly for constructed ideas, and thus the number of inputs may be very large, each particular feedback path is closed only if *all* of the points of possible opening are actually closed, i.e., only when the activations of all the suitable inputs exact coincide, which in the end may not be very probable. In this instance the action of the operator in the transition from consciousness to awareness cannot be described simply as directing awareness onto the idea but rather as first activating the correct set of inputs necessary for closing the circuit and for transforming in this way an immanent idea in potency into one in act. In other words, the directing of awareness arousal, or the lens, consists of a much more active and goal-conscious participation (since directly aimed at particular ideas or sets) in eidetic activity than seems on the basis of term used. We must consider that among the very many possibilities of closing complex and interrelated feedback circuits only one particular constellation of activated inputs is suitable for the particular reasoning in the course of development. Thus, the simple awareness-directing operator must generally, i.e., when it is only focusing upon a whole region, proceed further by trial and error, closing a circuit at one point and reading the idea, comparing and discarding it if unsuitable, closing the circuit at another point, reading the idea, and so on. All in all, this seems a rather clumsy and time-consuming procedure, and there should be faster procedures for performing the same function.

Awareness may operate in two different situations: a) when constructing an idea from other element ideas, i.e., when concentrating upon a single object only, and b) when developing thought, that is, when the ideas are taken one after the other and in some way connected, but each one is accepted without further perusal because its content is already known and unquestioned. However, in both procedures one may find ideas preferentially connected to each other by their close affinity or coherence —even if a variable one, according to the specific logic of the contingent thought. This leads us much farther than anticipated. Accepting that there may be preferential paths for thought, determined by some (casually) contingent logic, means accepting that thought itself may be constrained into a path, and that mind may have little or no part in the choice of this (and in some cases also no responsibility). In fact, if the ideas are closely connected to one another through reciprocal congruence and coherence (which may occur not only when mind in awareness has previously recognized a logical link, but also because the ideas belong to a preordered, closely connected region) thought *must* follow what appears as the more logical path. (The same considerations were advanced in Part I.) However, there may still be much more to what we have just said. The path of logic that thought must follow is the line of logic connecting the ideas all along the whole thought path from premises to conclusion, whereas the logic linking the ideas to one another is a reflection of the reciprocal congruence binding them, which generally is not changed when used in different thoughts (i.e., under different coherence lines). Therefore, the necessary conclusion is that for each particular line of thought, all along the reasoning path, one selects among the existing congruent ideas only those fitting that line's logic, leaving the other unused. However, to be able to perform this selection, the logic of thought must preexist, already preordered, and be present at the origin of thought. In other words, an entity (operator?) should be able to organize the particular reasoning on the basis of the ideas present and immanent and also to construct the right kind of ideas suited to that reasoning if they do not already exist in the idea space. Then, this entity must "know" the overall logical line of reasoning from beginning to conclusion even before consciousness is transformed into awareness. If this were not so, the development of reasoning would have a haphazard character because the broad system of inherent congruences among all the possible idea-elements of thought, could easily be mistaken for the intrinsic coherence needed in the particular reasoning. (We leave this point open until the discussion in following sections.)

The possibility that there may be preferred, if not altogether predetermined, paths of thought rests, on the one hand, on the existence of facilitated synapses and thus of anatomically and chemically defined pathways, and on the other hand, on strong congruence-coherence among ideas not to be overlooked by the exploring awareness. One cannot deny that the hypothesis of some anatomically and chemically facilitated paths among the maze of circuits embodying the ideas possesses a strong likelihood. Obviously, however, this would only allow an almost mechanistic development of thought, perhaps suited especially for coping with situations requiring an accurate appraisal of the conditions (possibly also considerably complex) and quick decisions since it allows only a very limited choice (or none at all) of alternate actions —decisions— in terms of reasoning. It seems very likely that such preferential, facilitated paths are not actually preformed from the origins but are formed in the long run as a kind of adaptation process (or learning) of the eidetic system to a repetition of situations or to situations of specific significance for the individual, requiring fast decision and fast behavioral response. Instead, we tend generally to accept that awareness is always able to select from the maze of eidetic circuits those exhibiting the correct type of coherence required by the line of reasoning. Only this procedure would grant thought all the freedom we feel necessary for developing in whatever direction.

The preexistence of an ordering logic of thought development does not eliminate the possibility —actually, the probability— of continuous interaction between the logical order of ideas in the state of simple consciousness and the logic of the ideas as realized in reasoning by the aware mind. Awareness of mind is a prerequisite for the critical test of reasoning by examining the overall logic through the filter of language. In fact, the perusal of contents at every single step may require reordering the thought. This is not a minor point since it entails that a) there may be differences between the preexisting and the realized logic; b) consequently one may be better (or more true?) than the other; and c) the preexisting logical pattern of development may only suggest possible lines, while the actual, final form of the thought is defined jointly by the preexisting logic and the logic of language of the aware mind (this would constitute the true and deeper sense of their interaction). These possibilities outline three aspects of the same problem.

Obviously, the question of a preexisting logic is closely associated with the nature of the ideas ordered by that type of logic. If a logic preexists, this can occur only in a space different from that of the aware brain, although communicating with the latter in a still unknown way but probably having the nature of a suitable two-way space transformation (or topological transposition). In fact, the essence of this space must consist of properties different from those of the space-time of the aware brain; otherwise the two spaces would coincide, which would mean that all possible logical lines of thought development are imbedded in the brain matter in the form of physical constraints (patterns of anatomical or functional synaptic arangements). It is only natural, instead, to see the space of the preexisting logic as that of the essences, of the ideas abstracted at the higher steps (levels) of refinement, since at their higher level they are ordered according to their essentially logical links —purely logical relationships— whereas the links among ideas embodied in the space time of the aware brain are ordered according to the variable space-time logic. In other words, the logic resides in the links. Thus, the two spaces must exibit a difference. The one that we consider as abstract phase space can only be represented by the immutable (time independent) ordered ensemble of higher abstracted ideas, a timeless region. From the latter originate all the ideas obtained by reversing the process of abstraction, from the ultimate essence to the particulars, to be embodied in the space time of the neural circuits, and this constitutes the realized, physical phase space. Also the transposition (or transformation) from one space to the other would assumes a somewhat different character from the other types of transitions. What the concerned operator must be able to perform to realize the "communication" between the two spaces is first to turn on the sequence of "downhill" steps from pure essence to particulars (an operation very akin to the analytical procedure discussed in earlier sections) and then to start the procedure of interaction with the time and matter of the implicit circuits of the aware mind. In doing this, the operator is confronted with the problem of topological language transposition between the two spaces.

9.4 Logic and Will

The operator seems unavoidably to be guided by will. The necessity of realizing a thought, i.e., an ordered sequence of ideas and propositions developing along some particular line up to a goal, is connected to a voluntary act of the (aware) mind. Thus, will proceeds in selecting from the abstract space an ordered ensemble of essences and gives rise, in turn, to the downhill procedure, the only one that can provide mind with the set of ideas necessary to start the reasoning in space time. Reasoning can proceed only in the awareness space time, by letting the original purpose of thought, included in its preordered logic, to interact with its realization in the brain matter. In this way, one is led from the space of preexisting logic —the space of the preexisting abstracted objects linked by their own logic— to the space of realized ideas (implemented in the brain circuits) also linked by their own logic. However, one must consider that the latter logic depends strongly on the characteristics of the physical system in which thought is implemented and is unavoidably therefore on a level different from that of the logic of the abstracted ideas, as it can represent only an imperfect transposition of the latter. Saying that the operator is guided by will may appear to be an overly committed statement. The nature of the operator, as well as of will, is still unknown (see, however, Sect. 10.3); thus, it is unwise to advance hypotheses based on this uncertainty. However, at this point we can make some speculations at least to simplify the picture. We have already encountered will as a major factor in developing thought, but we have not yet tried to delineate its properties. It should be clear by now that will as the promoter of a line of thought, can act only following a logical design. However, I would like to state this in a slightly different way: in eidetic operations will acts only through logic, which is only one step removed from asserting that will and logic coincide. In fact, at the level of promoter of ideation, i.e., of organizer of thought, logic may be considered to reside in will, i.e., the strength of the logical links may constitute the origin, the very nature of will as it manifests itself in the eidetic processes.

This shifts the question to the nature of will alone. In the brain space time of aware mind, will-function organizes the brain's activity toward selected goals; it can only be a "realized" will. However, to the realized will-function in space time must correspond the nonrealized will-logic in abstract space, that presenting the possible set of ideas to be realized in the space time of brain. The term "present" shrouds the true essence of will since the latter cannot be anything but the origin of the operation of transformation, or of transposition, from the possible constellation of logical links between the abstracted ideas of phase space to the realized constellation of logical links in the aware mind space time. Thus, in eidetic activity will has the nature of an operator and is structurally similar to the logic-operator linking the ideas to one another, inasmuch as it is coincidental with logic. From these arguments, one might consider will as the fundamental operator concerned with thought. The point is that will, if considered as the operator of transformation from one logical space to another, should assume the form of a tensor —such as the fundamental, metric tensor— transforming one into the other of two systems of relationships constituting the reference frameworks of two spaces (see also *PTN*). In turn, the nature of the operation itself may also be better defined, considering that will would operate the transformation from a reference abstract framework —part of a phase space where time is not involved, since the ideas are immutable— to an actual space framework with real, physical parameters, to which time also belongs. It seems awkward to endow a single "object" —will— with properties such as logic and operator of space transformation because, in some sense simultaneously logic and operator, will would be transforming itself from an abstract into a real state; rather, we should perhaps consider the two properties as facets of a compound single object. However, this autotransformation of will, which could only be autonomous, i.e., "free," would be an essential property of the *Ich* autodetermining itself (a concept developed by Hegel, see also Sect. 10.3). As has been noted above, what defines logic is the affinity between essences, the more abstracted at higher levels the more they become reciprocally close, i.e., the more they "attract" each other. At this point, logic and will may be identical, on the one hand, as will externalizing itself (Hegel), and on the other, as logic externalizing itself in eidetic activity through the transition from the logic of the timeless phase space into the logic of the brain space time. The "strength" of logic is the strength of attraction (the affinity of the objects) and constitutes the determinant of the voluntary motion; in other words, the voluntary eidetic motion may be initiated only in the form of a gradient in the strengths of the two logic-ordered plans of relationships, of the abstract and of the physical phase space. Fundamentally, at the base of the problem is the different nature of the two spaces, making it extremely difficult to envisage the transition between the two as a mere transformation, as is commonly understood in physical spaces, where the relationships between the parameters are subject to change, but the parameters as such always remain the same.

9.5 The Phase Space of Ideas: General Properties

We now consider more closely a number of problems related to ideas and eidetic functions. Considering only the processes in the aware mind, that is, in the actual brain space-time, according to our theory we view eidetic activity as bringing about a deformation of the brain space. (We maintain in *PTN* that the brain space-time has Gaussian properties, i.e., allowing different metrics in different regions, depending on the forces' fields —whatever their nature might be— present within that region of space.) Thus, also for the eidetic system we assume that there can be relativistic transformations of space and time (see also Sect. 2.3). The spatial coordinates of this space, ultimately representing ideas, exhibit different degrees of correlation to each other, a measure of the reciprocal congruence-coherence linking them. We mean here that sets of ideas, whether natural or complex (those containing general philosophical elements), can build subspaces with their own local metrics. Therefore, while in the whole brain space time the metrics may be defined, for instance, by different degrees of general alertness and involvement with the eidetic system, within the eidetic subsystem belonging to brain space-time we must postulate a much greater variability of local metrics, one for each subspace set of congruent-coherent ideas. However, the whole eidetic system itself, continually developing ideas and thoughts, may be viewed as variously spreading its influence over the whole brain, though very differently in its various subregions (as the areas in which atmospheric turbulence arises spread their influence over the world). This picture may represent the relationships between the eidetic system and other nervous system functions, expressed above in the statement that mind actually *is* the whole brain, a mind with sensors and control systems for bodily functions and for its own interactions with the outside world.

Let us consider the specific properties of the abstract eidetic space. In the physically brain-realized phase space the coordinates are Lagrangian generalized positions and momenta of the neurons embodying the ideas; thus, the coordinates refer directly to the ideas themselves, for they are each embodied through a number of such neurons suitably interconnected. However, the parameter time referring to position and momentum may be disregarded since the ideas are immutable (time independent, either as fixed in a stable, durable memory or, according to our theory, as continually renewed). The abstract phase space is differently structured at complete variance with the physical one; there are no neurons at the basis of position and momentum coordinates, and it enjoys a complete time independence since time does not at all fit in the abstract nature. We can say that in a physical phase space what matters are the real, actual values of the positions and momenta; in fact, these are the base for predicting the system's future. In abstract phase space, all the positions corresponding to the ideas can be considered as fixed once and for all, and while arbitrary initial values can be assigned to the position within the ordered structure of the position coordinates, the momenta —since the positions are fixed— can be defined by the stresses between the ideas, or by their "effects," i.e., the steady strains imposed by the stresses onto the ideal, fixed structure constituted by the ideas. As stated above, we can think of a force of attraction between ideas, a stress defined by the congruence coherence linking one to another. Among the ideas of the phase space, the time-independent cross-relationships outline subspaces in which the congruence and coherence among ideas are more strict, meaning that the contents form more homogeneous subgroups. We could say that each subspace thus outlined "gravitates" around a conceptual core, i.e., that the subspace is created by deformation due to a center of "attraction." As in the brain matter functionally homogeneous subsystems are created according to the needs, by the continuous adaptation of the organization of anatomical connections between the neurons and of their chemical properties, likewise in the abstract eidetic phase space the content (the meaning) has the role of an organizing factor which brings into existence homogeneous subspaces consisting in systems of correlated ideas (as blocks of ideas), the actual basis for organizing thought (see below).

Since the relationships among ideas relate inherently to their intrinsic contents, they are also immutable, i.e., fixed, constituting in this way the firm basis of the thought-organizing logic. However, if we were to say that in each region of idea phase space there is a local metrics, as was assumed for brain space time, we would be very wrong. In a space in which there is no time reference, the strong affinities among the fixed space elements can affect only the spatial coordinates, building stable patterns of lines of force resembling deformations, however excluding from them any time influence, meaning that relativistic phenomena cannot be observed. In fact, we cannot possibly speak of local metrics (the most characteristic feature of the Gaussian reference system as far as brain is concerned) in the precise sense until there is no "motion," and the parameter "time" is not involved. In the eidetic abstract phase space, the presence of subspaces and their related force fields together with the system of relationships among the subspaces (particularly strong when the latter consist of the so-called broadly philosophical concepts) constitute the fixed, time independent system that we have called the structure of the phase space where relativistic phenomena cannot occur, as in a physical space without motion.

Again at variance with physical space, however, even without motion the eidetic phase space is somehow "alive." As we shall see in the following sections, we meet here perhaps an indication of what scientists have hitherto left unexplored —at least within their usual scientific concern. Since it is abstract, this unknown is commonly taken as a matter only for philosophical notions of being and not an essential property of its essential, real nature. Of course, we wish to by no means imply that in the idea phase space "events" such as we observe in physical life may occur (every event necessarily develops in time and its ideal motion described as changes of Lagrangian position coordinates, thus, by definition extraneous to the nature of a phase space). The signs of "life" in the space of ideas can only be inherent in the particular nature of that phase space, in which the logical content constitutes the link among ideas and partakes of the abstract part of mind. Yet, no exhaustive description of abstract phase space results from this. Every line of reasoning requires new aggregations of ideas into other sets, and since these also represent so many systems of relationships among ideas, they are logically linked. The recognition of more abstract contents of ideas or sets necessarily brings about the transformation of subspaces or convergence of subspaces, etc. Nevertheless, all of these transformations or processes in the phase space are also time independent.

All the elements, ideas, fields, regions of ideas, in other words, every component of possible thought and reasoning is integral to the phase space, preexisting to every present thought and reasoning. Everything exists in the form of a structure made up of the logically defined, "necessary" relationships binding those components and thus constituting the "logic." (We leave aside the question of whether we are confronting an absolute, universal logic, independent of every mind or a personal logic, different for each individual.) Consequently, the preexisting components and their reciprocal, manifold relationships —i.e., the logic— constitute the basis of what we here call an "internal" language, which does not consist of the words or symbols used when the lines of thought are implemented in the brain circuits, and we are talking silently with ourselves (we think) or aloud with others. Since there are relationships, the connections among ideas and concepts are not in the form of words; they represent the constraints existing autonomously among abstract objects of mind and take the form of words only at the moment of their brain implementation. Thus the terms life, relation, transformation, and convergence, as referred to phase space, are misleading in conveying a seeming time dependence, and they are used here only for want of more precise terms. In fact, we must try to envisage *everything* as existing simultaneously and always present in a timeless space, all the possible origins of what in brain space time is creation, transformation, and convergence, i.e., the eidetic life. In this way, the phase space is understood as the seat of all possible logical combinations of elements and ideas, where at each moment and instantaneously —i.e., in no time— the preformed or preexisting logic has at disposal the suitable set of ideas for every need; we might say that the phase space presents them to the brain to be implemented (this is why it may be said to be alive). It is obvious then that in phase space the fundamental objects must be, on the one hand, sets of ideas, essences, concepts from which all ideas originate and, on the other, their reciprocal logical links. However, since the process of abstraction leads only progressively through higher levels to the essence of every idea, at the same time only progressively we are led to recognize the relationships among concept-essences of the same and of different fields. The point here is that for every level of abstraction there is an obvious, local logic linking the ideas of that level and of that level only; in other words, there exist different logical networks of connections, dependent upon the level of abstraction.

9.5.1 Ontogenesis of Phase Space

When and how is the phase space built? It is clear that the same aware-mind process which leads to full abstraction would also, figuratively speaking, open the door to the phase space of essence-ideas. Recognizing the result of a completed abstraction is possible only because we can compare it, or make it interact, with the ideas constituting the abstract phase space. Consequently, one might say that thought, once it has reached the stage of comparison with the components of phase space, becomes self-explanatory. Correcting errors of reasoning is the outcome of a critical procedure by which the logic of the brain-realized thought is compared with the logic of phase space; here the new ideas resulting from thought development in space time must be eventually "deposited." Therefore, accessing the phase space requires the reverse procedure of realizing the phase space content in the maze of the eidetic circuits.

At first consideration, constituting a phase space should be thought a long and painstaking procedure. If one could inherit ideas as one does somatic characteristics, small children could exhibit earlier the maturity of thought and behavior normally acquired only in adult life after years of learning and training. That this does not occur means that in the beginning ideas are assembled and developed slowly. It is true that teaching is a most powerful procedure for developing the phase space, but when we teach a child that this is a dog, as is the one over there and the one farther away, what we are actually conveying to him is the confusing information that differently shaped individuals may have the same name —not the idea of dog as an autonomous entity. On the contrary, the idea dog, representing a class of individuals, is formed in the child's mind only when he himself is able to discover (to extract) an hidden essence identifying dogs from the manifold forms exhibited by the many types of animals given this name. We have already said that the power of abstraction is acquired slowly, and that training and learning through a teacher can enhance it. We can now add, that the long period required for maturation of the human mind is the time of growth of the phase space and of the hidden logic constituting its nature, the time of the abstraction procedures applied one by one to the different objects with which mind is concerned in the course of years.

The time required for forming natural ideas is also long. In the instance of the idea of dog, the child must first feel the illogicality of using the same name for different physical objects, when at the same time he is taught (as another instance) that mother is a name for only one unique object, although there seem to be as many would-be mothers as there are dogs. The search for logicality, or the feeling of an apparent lack of it, may well be the factor activating the abstraction process. However, the time required is much longer for ideas without a physical (perceptible) basis, the constructed ideas, due to their higher degree of complexity. Since there is no physical counterpart, their assembly takes longer, and teaching is needed to shorten this time. However, all the ideas concerning the broadly philosophical concepts —ethics, religion, philosophy, etc.— even without teaching can be constructed autonomously, though certainly more painstakingly. Therefore, they are expected to attain maturation in the aging individual —one might say that maturity of the individual is the maturity of his constructed ideas, certainly not of the natural ideas such as dog, house, or even atom, etc. Hence, correct behavior is also strictly correlated with maturity of constructed ideas. It is therefore no wonder that humans exhibit large individual differences, since social conditions may differ substantially, and teaching and experiencing may strongly vary across the social groups. Nevertheless, it must be stressed that in eidetic processes teaching —as the enzymes in biochemistry— can only accelerate a process, while the intrinsic capacity of each individual mind is left alone to confront the task of elaborating the ideas belonging to the concepts, in other words, the concepts' development (and this is another major origin of the differentiation among individuals.) In the end, we should expect the constitution of phase space to represent the result of successive transformations of the eidetic cerebral space time, though now complying to the order of the preexisting ordering logic, which cannot be learned or otherwise acquired as it is inherent in the essences of the abstract objects. The actual constitution in ideas of the phase space may easily differ in different individuals, since it is conditioned not only by teaching and experience but also by inherited factors related to family and other higher order groups.

Obviously, the development of phase space depends in practice on the ability to construct complex thought in an autonomous way, a property peculiar to man and an index of his true nature. As is suggested in *PTN*, man's intellectual capacities are determined by the complex structural organization of his brain, enabling him first to judge about problems and situations and then to behave according to human standards on all occasions. Therefore, a "human" quality of the logic is implicit. On the other hand, the concepts and ideas have reciprocal inherent congruence and coherence, which is another way of saying that there is a natural order among them, a natural logic (in the same way as there is an order and a logic in the nature that the natural ideas represent). The point here is that the structural organization of the brain can only set limits to the complexity of ideas and relationships (between the ideas) which the brain can detect, i.e., understand. Certainly we can be confident that the simple abstract objects of nature, for instance, the idea of house, in the sense of a suitable and convenient shelter, is shared by man and even dog with basically similar content. However the characteristically human way of conceiving of a house depends on the structural organization of his brain, and this invests it with a host of connotations not shared by any other living being. Nevertheless, also when objects of nature (of natural sciences) are concerned, the human brain's ability to understand the complex relationships among them may be strained. Still, we always are in a field in which the concept-ideas are abstractly symbolic, physical realities and find their logical explanation in the logical relationships between the objects and the events of nature.

The situation is altogether different with ideas from broadly philosophical fields. The inherent congruence among the elements and the intrinsic coherence among ideas —presented immediately to understanding when they have a natural origin— are not here available with any comparable immediacy. Each concept is translated into ideas rather being the result of a long (and sometimes tedious) internal dialogue of mind with itself. No comparison with natural objects is available, and thus no comparison with the logic immanent in nature; the justification of each idea and each thought must therefore be based on a logic that mind has either built for itself or recognizes as already existing. However, the criteria for assessing the logicality of connections among the components of thought —the basis of aware mind's own logic in brain space-time— obviously cannot come out of the blue, as it were; neither can it be determined arbitrarily, for contradiction would quickly emerge. Therefore the mind's first empirically outlined criteria themselves should be tested against other terms of comparison, i.e., against an already consolidated, logical system of relationships such as can be found in the abstract phase space alone. With this we reach the point at which our judgement must choose between two alternatives: a) mind itself alone builds its phase space logic uniquely by comparison or analogy with natural logic after suitable abstraction of natural concept-ideas until its system is completed; b) as part of the brain's structural organization, genetically defined as characteristically human, the bases for logical development of thought are also present —in this way also genetically ensured— not of course in the form of particular thoughts but as heuristic principles of thought development.

According to the first hypothesis, through an abstraction process, natural ideas and the mind's own constructed ideas can be compared, and the logic of the essences formerly derived from natural objects sustained by natural logic (the logic of the relationships among the elements of nature) may be taken as the logic at the base of broadly philosophical ideas (and thought) as well. No difference can result between phase space and brain space-time; the one must be the perfect, isomorphic transposition (as a topological, isomorphic mapping) of the other because the natural logic, to which the brain's logic is made to conform, is itself derived from the relationships that the brain discovers among the objects of nature. Since the ability to discover relationships among the objects and ideas in nature is limited by the structural organization of the brain, the brain's logic is self-explanatory at all levels (this allows only a perfect identity of content of the phase space and brain space time).

The second hypothesis posits that the structural and functional organization of human brain, in the part genetically defined, also contains the nature-independent, humanly-specific, fundamentally autonomous principles of logic which develops all types of thought —thus, not simply the ability to recognize natural relationships but for autonomously constructing nonnatural ideas (and thought). This means that an explanation of the brain's realized logic is not be sought in the confrontation with nature's logic but rather with an autonomous term of comparison. Again, this can consist only of the logic inherent in the relationships among the essence-ideas, but, and this is the point, also in those relationships the brain in its limited capacities is not able alone to understand fully or to explore sufficiently, although existing in the phase space of the essence-ideas and founded on the inherited principles of the heuristic development of logical thought. (By analogy, as in nature the discovery of atoms and electrons did not at first reveal all the relationships between these objects, likewise in a more abstract domain the relationships among such objects, for instance, good and society, are not all immediately, and perhaps, never completely, understandable, although in both physical and abstract domains the relationships are present from the moment that the objects themselves are present, preexisting to every thought development.) Because principles are embodied in the inherited structural organization, they are not elaborated in consciousness-awareness and cannot be recognized or experienced directly in phase space as sets of principles filed therein, but only indirectly, mediated through brain embodiment when the phase space presents ideas to consciousness-awareness. These principles are self-explanatory in the abstract phase space itself alone, and the logic of the brain-embodied phase space can only partly reflect the logic inherent in abstract phase space due to the inherent limitations of brain matter. Accordingly, the brain-embodied phase space cannot represent an isomorphic topological mapping of the abstract phase space, and must be explained by the latter. In the space of the abstractions, the relationships (correlations, congruence, affinities) are inherent in the ideas' contents; thus by definition these relationships in themselves cannot change, but at the same time they cannot be transposed into the brain space-time if the structural organization of the relevant eidetic circuits is not conformed accordingly. This is the source of the difference between the abstract and the physically realized phase spaces.

The acceptance of either of the two hypotheses may be based more on metaphysical considerations than on experimentally determined facts. Both hypotheses, however, may eventually find support in common neurophysiological data. It is suggested in *PTN* that the ongoing activity of the brain cells in steady state should exhibit stable patterns of distribution over different neuronal ensembles. This is taken as a kind of basic "tension", an outward sign of the transposition in the brain space-time of the abstract properties of man's mind (see the "Appendix" of *PTN*), some sort of map of physical constraints topologically isomorphic, though incompletely, with the map of logical constraints in abstract phase space. The two hypotheses do not even necessarily exclude each other at all points. Both are founded on the abstraction procedure, necessarily performed through a neurophysiological operation, certainly the most outstanding and peculiar property of human mind and at the same time probably a built-in capacity and drive to use it, although needing continuous training and practice. Abstraction alone enables man to approach and understand not only physical nature but especially the nonphysical, less accessible spiritual world.

10 Time-Unrelated Regions

At this point the question is not that as to the existence of a phase space consisting of ideas and concepts but rather about its structure and rationale. In approaching the core of the problem, however, one immediately becomes aware of the hidden question concerning another entity also to be accounted for, i.e., language. The nature, structure, and rationale of abstract phase space compell us to wonder about the way in which the relationships among the ideas and concepts constituting it are expressed. The term "express" is certainly misleading in this context because it conveys the notion of language in the way in which we are accustumed to conceiving it, i.e., as consisting of words organized into sentences. However, the essence of language in phase space lies in the links between ideas, that is, their essential relationships, not between the words used for expressing thoughts. While these links in everyday language used in thinking, talking, and writing are themselves symbolized in words, in abstract phase space they are only inherent in that which keeps the elements of phase space congruent and coherent among themselves, i.e., in the logic linking the abstractions to one another. This logic is the true origin of language and thus of the eidetic function which, from this point of view, is the same thing as the former. However, if the properties of abstract phase space are those proposed in Chap. 9, with properties different from those of the physical phase space (as embodied in brain space-time matter), then the language in the two spaces must also be different. In other words, the logic of the two spaces, although representing the principles of congruence-coherence linking their respective elements, must differ because the structure and the rationale of the two spaces are different.

10.1 Consciousness and Conscience

We should consider here first the meanings that we have attribute to the terms consciousness and conscience. Consciousness may be a state of mind or of the brain as a whole. One may be conscious of seeing, of hearing, of the presence of the world around, and of being alive without paying attention to particular items, so that what we may call a state of simple consciousness implies only being conscious of being conscious in general. In realizing the contact with the outside world, consciousness is necessarily bound to brain's space-time; however, it also is an abstract structure and a function. Consciousness, on the one hand, is the simple idea of the system of relationships linking the brain to the self and to the world, making us "conscious;" on the other, it is the embodiment of this system of links, not as the neurological function of sensation or perception but as the matter-bound realization of the principle superseding mind's and psyche's contacts with the world. Thus, simple consciousness appears to be a rather peculiar "object" since if the idea of it were not embodied and thereby made real, we could not be conscious; that is, the very realization of the idea in the brain circuits makes possible the realization of the state of consciousness and of the function of closing the contact between the idea phase space and mind-psyche, on the one hand, and between the latter and the world, on the other. But, it is immediately apparent that for this to occur, the state of consciousness must already be present, for otherwise the idea could not be embodied in the brain matter; therefore, idea and state of consciousness must appear (be present) simultaneously. Accordingly, although as abstract structure it does not belong to space-time, consciousness as state and function cannot exist outside space-time. Moreover, as already stressed, there is in the state of simple consciousness an immediacy and a simultaneity of presence throughout the brain structures as soon as the idea is embodied —which can hardly be reconciled with the usual speed of the brain functions. These develop, rather, only by means of nerve impulses along the organized maze of nerve fibers and through the chemical synaptic contacts on the cells. The brain's structural organization on these bases does not involve simultaneity of nerve events any more than general relativity does in physical space. Thus, simple consciousness, in order to justify its own characteristic immediacy, i.e., its being simultaneously conscious of itself (read: of the whole of its brain) and of the general existence of the world around, must rely upon operations proceeding at a speed inconceivably high for the nervous system. The relationship between consciousness and time is to be seen within the ever-changing and "timeless" present (see below).

The distinction which we now want to make between conscience and consciousness is that conscience is a function only of phase space —leaving aside further possible specifications of other essential and questionable, metaphysical meanings. In analogy with physical space, it could be said that conscience is *the* phase space of all ideas, or, rather, as an exclusively abstract structure, that it represents the "idea of all the ideas." Although an idea, conscience as phase space also plays a role, but it does not need to be embodied in the brain circuits to be operative, as simple consciousness does. Conscience as a role, if we accept its existence, is a peculiar object because it enjoys a freedom from time, consisting of the timeless region made up of the abstract objects simultaneously existing all together. For this reason, no events are allowed in this region, since every event would necessarily be bound to a time development, to a before and an after. Nevertheless, we may say that conscience constantly presents its blocks of logically ordered concept-ideas for transposition into consciousness. However, as we shall see, if one does not accept the existence of such real, nonphysical objects but only matter-bound objects and exclusively physically based hypotheses, conscience must also rely upon a physical structure consisting of all the eidetic circuits embodying all the ideas available and eventually coincides with the physical, matter-bound phase space discussed in Sect. 9.5.

The reciprocal relationship between conscience and simple consciousness is certainly very difficult to understand. That this contact exists is shown by the fact that the ideas implemented in brain circuits must match in full or in part the ideas of the phase space. Since both conscience and simple consciousness are real, and yet on an abstract plane, they also belong to the same space, in which they may "communicate" with each other, as it were, in the sense that their abstract natures are so similar that they come to touch each other (we may say that they merge into each other). At the same time, simple consciousness, as a state of brain, certainly also exists in matter, and through the latter it partakes of the space-time constraints of the nervous system. This means that if conscience and simple consciousness communicate, they can do so only through the abstract properties of their common nature; also, only through the mediation of simple consciousness can conscience communicate with brain space-time. Thus, the world of phase space immateriality and the world of brain space-time matter cannot interact with each other directly but only through the unique entity having the properties of both immateriality and matter, i.e., the sole entity that can bridge the gap between the two opposite planes. (There is a profound significance in this. For ancient philosophers and medieval theologians this bridge function, or property, is peculiar to man, who is considered to be such a mediator, since his *anima* has properties common, on the one side, with those of purely material and, on the other, with those of purely spiritual entities; see: *PTN*, "Appendix.") We see, at the moment, no other logical way to explain how two different worlds may come into contact and interact (communicate) with each other. Without the bridge supplied by simple consciousness, the two realities, the one of the abstract space of ideas and that of physical space-time matter, would remain separate worlds. However, although this interpretation may appear altogether logical, its acceptance may still prove very awkward for the positivistic natural scientist who would not accept the existence of a real though abstract phase space in the absence of any type of practical (to this writer, impossible) demonstration, i.e., one that does not rest exclusively on the persuasive strength of reasoning's logic.

It is easy to see that conscience, as an abstract phase space, would have all the properties of memory; this would be of a very special quality, indeed, since there is no matter or eidetic circuit in which the memory could be continuously renewed, and there would therefore be no stored items although the constituents, i.e., the abstract objects of that space, would be continuously present. One must never forget that in nonphysical phase space the abstract objects cannot have material support —pertaining exclusively to the embodiment of ideas in the neuronal circuits, i.e, to the physical space-time of the brain. At the stage of simple embodiment, only consciousness exists —better, this stage *is* consciousness, according to our definition, concerned with the more general relationships of brain with its self and the outside world. The fact that simple consciousness is realized in brain matter (as idea embodied in brain circuits and simultaneously as state) and as such operates the bridgelike transfer between conscience and awareness means that through it the links of conscience with brain matter are extremely close also in the physical sense. Although conscience phase space cannot be given physical dimensions, and therefore also no seat in the brain, it must nevertheless be present precisely when and where consciousness appears, i.e., at the level of the neuronal circuits in which consciousness becomes a real, matter-bound function. The presence of conscience, at the moment that consciousness is realized, is revealed by that perception of speed and simultaneity of presence, such that all the ideas appear as soon as consciousness is activated, a property strongly differentiating the operations of consciousness from those of awareness (although both belong to space time). The peculiarity of awareness —what we termed above the "lens" of consciousness— is to make it possible to analyze individual abstract objects; these are introduced into the flux of events of space-time, ordered in thought, and compared to others, etc., all functions developing with the characteristic slowness of anatomical and chemical connections.

For students of positivistic, empirical sciences accustomed to dealing only with matter and its attributes and properties, it is understandably awkward to envisage the phase space of abstract objects as a real entity. One is accustomed to considering ideas as physical "products" of our brain, that is, something that brain may create or not, precisely in the sense that only brain matter conceives the ideas and gives them a perceptible shape, analyzes and organizes them in thought, etc. However, according to our conclusions, without a phase space from which the ideas are transposed into brain matter there could be for man no autonomous eidetic function, or memory. As suggested in Chap. 9, abstract phase space is richer in essential contents (i.e., in relationships among ideas) also in those relationships mind, in brain space-time, is unable to understand and therefore to embody (since embodying and understanding are the same thing) and finally to fully explore; these nonetheless exist independently as immutably inherent relationships among abstract objects. Only after suitable abstraction can mind also extract some of those hidden relationships, but for this to occur abstract phase space must present to mind the suitable blocks of ideas. Without these, mind is cut off from the wealthiest source of thought items. Thus, if one looks at the problem a bit closer, one should be less concerned with the possibility of retrieving ideas from phase space —which, of course, remains a subject for investigation— than with how the ideas get into it in such a way that they can preserve the fullness of their contents, in other words, the way in which this type of phase space comes into existence. However, the core of the problem is still in the real essence of phase space, i.e., in what it represents.

Let us examine this question more thoroughly. Suppose we start with the newborn's brain. We assume, although it is not proven, that at such an early phase of life there are no ideas in the brain (no inherited ideas, at least not the symbolic abstract structures of concepts), and therefore that everything is acquired through experience —beginning with natural ideas— and particularly through learning from the parents. One may wonder at the extremely long time required by children to attain mental maturity; however, this long period is justified if one considers the nature of the task that the child faces: developing the ability to extract essentials (i.e., that which does not immediately affect the senses) first from the objects of nature, and then from the objects of successive steps of abstraction. Once the procedure is learned, mind may proceed autonomously with ideas no longer related directly to natural but to ever more abstract objects and finds the essential relationships among them. This process involves "building knowledge." Thus, knowledge itself is what results from those transformations operated by mind on the objects of nature and other ideas. As such, it constitutes a first phase space of ideas, made up of the natural ideas that the aware mind finds by contemplating the objects of nature, the ideas at somewhat higher abstract levels, and then of the ideas found by recognizing their relationships. Thereby, the disorganized collection of ideas takes the shape of a system, i.e., becomes "Science" (see "Introduction", this volume").

However, the picture is still not satisfactory. The list of ideas of the first phase space lacks constructed ideas because to be constructed these need a "ghost" of preknowledge, something that mind by itself is unable to build (see Sects. 8.1, 8.4). We can now be more precise in saying that the preknowledge contained in this ghost belongs to a concept. This is a nonsymbolic abstract structure (although retrospectively we may symbolize it with an arbitrary or a general name) imposing itself for immediate understanding —well before mind can give it more precise contours— as the sole self-explanatory reality. The full richness of contents is comprised from the very beginning in concepts, and from them mind derives a host of ideas. Let us take, as an example, the concept of justice. As soon as our previously innocent person suffers the first time from an evil or unpleasant experience on the part of another person, he is struck by the inner feeling of a wrong. In him is thus born the concept of justice, certainly in a vague form, when his language ability is still very undeveloped. Of course, the concept could not have the richness of sophisticated contents of justice. The same, now less innocent person then develops the concept in the form of ever more clearly specified ideas. This concept is not clear (not fully defined) at the moment of its appearance; the relationships among all the ideas that derive from it remain imperfectly explored now even after thousands of years, and for each person they must be approved by the concept that each one has in himself since he first suffered injustice. However, and here is the point, it would be incorrect to say that the concept was subsumed within an abstract phase space as one of its first items. The concept itself has the structure of a phase space; actually, each one *is* itself a subordinate phase space made up of all the relationships among its respective elements-ideas, and since it encompassess all of them, it may give rise to all the possible ideas made up of those elements. Thus, the phase space first consists of the ensemble of all the concepts that flashed to mind and then also of the derived ideas, but only because these represent systems of relationships among the elements together giving an understandable content to each of the concepts. Accepting that the phase space is made up primarily of concepts (at a very early step even the very presence of ideas may be questioned) accounts for the difference between the phase space and the whole set of ideas embodied in matter, as the concepts are a potentially inexhaustible source of elements, of which only a part can actually be embodied in matter in the form of ideas (as relationships among elements of the concept).

As we have said, the phase space is made up of logical constructions, every concept constituting the system of all the existing relationships among the relevant elements. The ideas with which mind clothes the concept define the limits of the powers of mind. Mind's limitations rest, on the one hand, on the absolute finiteness of brain matter and, on the other, on the precision of the transposition procedure from phase space to space-time; both are heavily conditioned by the basic possibilities in terms of material structural organization of the eidetic circuits. It is certainly safe to assume that in every concept, as a primarily autonomous entity, there exist relationships among elements that escape the ability of mind to elaborate corresponding ideas, i.e., in every concept there are things (constituent elements and relationships among them) that mind is unable to grasp and completely develop. The same procedure by which ideas are formed, through the recognition of inherent congruence among elements, and the procedure of thought development through recognition of intrinsic coherence among ideas force us to accept that there may be relationships among the elements as part of the ideas and among the ideas as part of thought, escaping the capacity of the mind which examines them in a state of awareness. As we have said, these relationships, however, really exist —although they remain hidden and unnoticed in the phase space— and are comprised in the concepts from which the ideas are derived; in this space they build a more complete system of logical relationships than may be comprised in the brain matter. The phase space must therefore be considered the unique, complete logical structure, the ensemble of concepts and derived ideas. The relationships among element-ideas cannot all originate from mind itself since they could never be sufficiently explored by it.

10.1.1 The Concept

We have said that upon close perusal the real question centers on how the phase space comes into existence. We have just concluded that the phase space is first built up of concepts. This seems to shift all the issue onto the concept itself, not only as to its nature but also its way of being in the brain and the phase space. Saying that every concept has the structure of a phase space focuses the question on the meaning of the sentence above: "the feeling struck his mind" (of the innocent child) in an almost flashlike way. With this, we stated that mind does not construct the concept but only ideas as would-be descriptions of a concept. This is thus a more comprehensive object than the idea and is the origin of a number of them. In this way the concept imposes itself as the fundamental abstract object, not created by mind but discovered piecemeal by it. Viewed from this perspective, the concept possesses a reality of its own —in agreement with Hegel's view that only the concept has a reality, and that the concept gives that reality to itself. Only a concept can be the first object grasped by a not yet reasoning mind, in the sense of a mind that has not yet developed the ability of symbolic language. Obviously, that inexpressed object is full of still obscure contents, although capable of guiding man's acts in a way that sensory experience alone could not do (and independently of it), and eventually of supporting language and thought as a basis for the newly acquired ways of behaving. In this sense we also meet the definition of the concept given by Thomas Aquinas: *mentis interius verbum*, the inner language of mind, not needing a *sensibile signum*, a perceptible sign to be understood, i.e., not needing the symbolic language used by mind for thought. However, when we say that mind grasps a concept, we are still very imprecise: the "grasping" here is hardly more than a feeling, a still hazy consciousness of an "existent" not belonging to the materiality of the senses' realm and yet from the very beginning being the reference for that which the senses tell the brain and mind further elaborates.

When mind develops symbolic language —the "external" language— it gives a name to every concept and to ideas. In this way mind may start "thinking," however, no longer using the internal language (which persists) but the new one, much more precise and able to describe the individual contents contained in the concept, although much less comprehensive than the latter as it is incapable of describing all the hidden meanings of it. Still, concepts and derived ideas, although named and described, are the abstract objects making up the structure of phase space, and if we regard the things a bit closer, we see no reason why we should say that "mind grasps abstract objects" rather than "abstract objects grasp mind," or better, "mind is all abstract objects," i.e., "mind is phase space." Thus, mind phase space has at least two velocities. When we say, incorrectly, that mind "talks" to itself through concept-ideas of abstract phase space (the internal language), mind operates rather instantaneously, with the immediacy of a quasi-infinite velocity; however, when it speaks with symbolic language, mind can only work with the slowness of the neural structures in which the symbolic language is embodied. The point is that the concepts may exist for us (in the sense that we may "feel" them, if imperfectly) only in the state of consciousness, i.e., when the abstract phase space comes to partake (we must say, in a still very obscure way) of the materiality of nerve activity. The problem of human communication independent of perceptible signs should be seriously taken into consideration together with the transition mechanisms from conscience to consciousness. In the state of consciousness, the materialized mind phase space comes into instantaneous existence in the brain; mind *is* the brain, as was said above. When thought is developed in awareness, on the other hand, the brain space-time procedures are used, and the velocity of mind operations is subject to space-time laws. We can now simplify our arguments by accepting that the imperfectly materialized mind phase space may after all be identified with simple consciousness, while its counterpart —fully abstract although in all senses real and the only one having a complete reality by virtue of its conceptual essence— is conscience.

10.2 The Timeless Region

Let us now return to the questions associated with the parameter of time in eidetic spaces. We have maintained that abstract phase space, on the one hand, and brain physical space, on the other, have different relationships to time. The phase space of concept-ideas, representing the immutable essence of things, is itself also immutable; time is not a necessary coordinate of this space since the concept-ideas remain what they are. In this region a "before" and "after" cannot exist; everything is there present and immutable, allowing only development in the number of constituents —a process sharing nothing with time-dependent development. We could say that this region is outside the influence of time since it is not subject to anything resembling development. These properties lead us logically to the immanence of the elements and concept-ideas constituting the phase space. Once more we must stress that the phase space is not the "container" for concepts and ideas but *is* the concepts and the ideas together with all their reciprocal relationships considered at all levels of abstraction; the concept-ideas themselves have the structure of phase space. When we consider the materialized phase space (the state of simple consciousness) by accepting its real existence, we cannot also say that it is outside time for the very reason that it exists in matter. "Timeless" in this case can only be meant as being always the same, a state of being not very different from a steady state at least as it partakes of nerve cell life. In the general case of a steady state, one recognizes the presence of determinant forces for all ongoing processes, a steady state being reached when the force effects are balanced or the velocities of the processes attain constant values. A system in steady state maintains continual "exchanges" with the other systems, e.g., of material constituents and energy, while the measurements of its parameters maintain a constancy over time (a situation certainly deserving to be accurately explored in terms of simple consciousness).

The space-time of the brain encompasses all the brain functions in general, i.e., all the cerebral activities developing in the space of neuronal circuits. In this region, the events are causally connected, and time is a necessary parameter when there are processes, i.e., evolution. It is the space in which past, present, and future have an exact meaning, there being a here and there, a before, a now, and a later. In the space-time of brain every motion is subject to the laws of a limit velocity (although, not necessarily the velocity of light, but a brain counterpart to it). Accepting this framework, in which relativity may play a substantial role, we implicitly accept as well that, by analogy with physical space-time, brain space-time events may also be envisaged as being ordered in so-called "light cones." This is the spatial representation of all the world lines describing the motion in space and time of the past events, forming a cone converging at the apex in the infinitesimal point "present here and now" and diverging into the cone of the future events. The limiting surfaces of the cones are defined by the events occurring at light velocity, the limit velocity of the universe. Anything outside the light cones (if ever anything can really exist outside them) is indeterminate and indeterminable since it could not be explored with physical means, ultralight velocities being required. Whereas, when the characteristic limiting velocity of the events is well below the velocity of light, as in the case of the neural events, the world lines are all inside the universe's light cones (the light cones of physical space-time) and representing a subspace of them.

The space of concepts-ideas, where one cannot speak of before or after, cannot logically be inside the physical light cone. Rather, because of their immanence without reference to time, the concept-ideas may be imagined as constituting the indefinite, timeless region outside the light cones of physical space-time. The continually existing phase space may, however, communicate with physical space-time, i.e., it can become "real", and this can occur only at the point of convergence of all the world lines, the infinitely small "present here and now" at which the past light cone transmutes into the future light cone —the only point at which the constituent of the phase space may be thought of as entering the stream of brain time events. According to mathematicians and physicists, some suitable transformation of coordinates may ideally make all the points of the region outside the cones simultaneous with the unique point "present here and now." Even if this is a mere mathematical possibility, since no "physical" causality could be produced by this in Minkowskian space-time, the matter still seem worthy of investigation as far as the eidetic phase spaces —material and immaterial— are concerned. In the same way, considering a reversed transformation of coordinates (if mathematically possible and physically meaningful), we can envisage how these past brain events, which have assembled elements into ideas embodied in the brain circuits, might communicate with the region outside the cones —by ejecting themselves into a timeless existence— that is, how the abstract, eidetic phase space might be developed through the door of the present.

From this perspective, conscience and consciousness acquire better definition through their different relationships with the physical light cones. A Maxwell's demon hypothetically exploring the region outside the cones would be able to encompass it with infinite speed; we might call it an "instantaneous" explorer. This we described as a property of conscience: encompassing the whole world of ideas —the whole of knowledge— simultaneously and instantaneously, as if scanning it at infinite velocity. This is a highly improper eventuality since conscience itself constitutes the phase space and thus, properly speaking, does not encompass itself but simply knows itself —better, is "self-knowledge," i.e., *is* knowledge. The moment of contact between the region outside of time —abstract phase space of ideas— and the present at the converging point of the light cones —physical space-time— is the moment when consciousness instantaneously and simultaneously takes possession of the brain matter and, at the same time, the content of conscience; the whole procedure is the result of a transformation of coordinates. From that moment on, the speed of the hypothetical explorer must slow down, as demonstrated by the fact that we may begin to observe "events" as they occur exclusively in the here and now of the physical space-time. In the state of simple consciousness mind would still make no reference to time actually flowing, since a time reference implies with the present also the past and the future. As the state of consciousness of "frozen presences" (of things inalterably present), it should be considered as the more faithful (isomorphic) transposition into brain matter of the timeless phase space. The difference between the two spaces lies in the fact that simple consciousness cannot be the exact image of conscience (abstract phase space) since the limitations inherent in brain matter make its space-time processes unsuitable for fully comprising and thus for fully understanding the (content of) conscience. Limited capacities of matter means that limits are imposed on it by the state of its constituting elements; in other words, since matter is spontaneously undergoing unceasing motion, the limitations stem not only from the functional organization of the nervous system but also from the physico-chemical processes which keep it alive.

For those who want to consider ideas as "exclusively" spatial and temporal events requiring finite times, i.e., events solely of matter and time space, the state of consciousness would also encompass them all almost at the same time by virtue of a very high velocity, that limit-velocity motion defining the surface of the event cones of the nervous system (i.e., of a subspace of the light cone of physical space-time). This velocity could still be very high but not necessarily only in comparison with the standard velocities of all other brain events. The latter interpretation is of paramount importance because by eliminating from the picture of eidetic organization in our theory the utterly unknown region outside the physical space-time cones and its hypothetical relationships with events inside the cones, it is possible to explain with exclusively physical means the interplay between a physical (matter-bound) phase space and the rest of brain matter, since the events of the former would occur at a velocity lower than that of light. However, by eliminating the nonphysical phase space of abstract objects, we would be confronted with the need to allot also to concepts a stable physical seat in the brain matter in which they could remain immanent, be retrieved used, and all this without a clearly defined form, either abstract or physical. In fact, the so-called physical phase space would constitute a stable memory of something completely indefinite, and moreover subject to all the considerations advanced in previous sections. Before reconsidering the questions associated with the memory aspect of a purely physical phase space, however, we must discuss further other aspects connected with the hypothesis of the physically dimensionless phase space of abstract objects.

10.3 Overview of Relationships Between Will and Eidetic Function

In the preceding section we examined the relationships between time and the two regions of ideas, the one outside the event cones, the so-called "timeless" phase space of ideas, and the other inside the event cones, the brain space-time. The latter obviously needs no further comment because eidetic activity, when embodied in the brain matter is also bound to the time of all the nervous processes. (In this instance one must always account for the hypothesis advanced in *PTN* that the existence of a "personal" time suggests the possibility of relativistic deformations of brain space-time.) The presumed real existence of the phase space as a region outside the event cones, i.e., not physically (in the sense of causally) connected with physical space-time inside the cones, raises difficulties in interpreting and understanding the situation. On the one hand, the existence of this phase space as real entity and not simply a mathematical concept challenges the mind; on the other, the mechanisms of the necessary reciprocal interaction between the two regions remains —perhaps forever— a nondemonstrable assumption.

Certainly we meet semantic problems when exploring the border region between physics and metaphysics. The terms conscience and consciousness are commonly used in a number of imprecise and ambiguous ways: conscience, in the areas of religion, ethics, and philosophy is distinct from the "state of conscience," i.e., consciousness; on the other hand, the two terms are generally confused in neurosciences. I did not want to introduce new terms to explain further issues in this topic. Rather, I will try to make more clear their various implications (as in the preceding sections) and their reciprocal relationships. Despite the wide use of the two terms, very few attempts have been made to establish the connection between them. Neurophysiologists do not attribute to conscience any scientifically accepted meaning beyond "state of conscience" while they attribute to consciousness a number of different and not universally accepted meanings having to do in some way with awareness.

Following the common meaning of the term, consciousness for the neuroscientist usually means a "state of attention," a meaning that may be enlarged to include "guidance," and this on sound logical bases. Actually, the meaning of attention is rather clear although the underlying neurophysiological procedures are far from clearly understood. Attention has always referred to objects belonging to some specific and circumscribed field (such as objects of sight, of hearing, of motion, etc.). Accordingly, the neurophysiologist has never considered the undifferentiated, or simple, aspect of consciousness, that of being conscious in a general sense, not to any particular item or field but simultaneously to all of them and therefore to none in particular. In this way, simple consciousness as it has been simply understood here is the state of being conscious that one exists, the self-consciousness of "being there." In this more general sense it surely does not bear any resemblance to the other meaning attached to it, guidance. When consciousness is understood as attention, it is directed (or directs itself) to specific fields or items; then consciousness and guidance —in the sense of directed or directing consciousness— tend to coincide. However, when consciousness is given such a meaning, it becomes the same as "awareness," the term widely used in neurosciences, a consciousness at a more advanced stage than the mere simple one, a specificity having appeared and forcing it to focus upon something very specific, either a field or an item (as above with the "lens of consciousness").

The guidance aspect of consciousness stands out clearly in eidetic procedures. Whenever during reasoning the line of thought is addressed in any desired direction, one must accept that awareness actually "directs" reasoning. It is apparent that awareness, as consciousness before, should also be considered a manifold object: an idea, a state of brain, and a function at the same time. The reason obviously is that consciousness and awareness are two different aspects of the same object, only appearing at different moments and with different purposes but each always easily convertible into the other. From this perspective, it is worthwhile to explore the hypothesis that the brain can be almost simultaneously in the two states, consciousness and awareness, the former always lasting for the infinitesimal time fraction sufficient for the appearance of ideas in the brain circuits, the latter for as long as mind is engrossed in reasoning. Particularly when constructing ideas from elements apparently heterogeneous and searching for their affinity by voluntarily refining the process of abstraction, the guiding character of awareness is most evident. Thus, the two aspects, attention and guidance, are linked by the necessarily voluntary character of the operations of aware consciousness. As we have just explained, only when attention is not focused —not self-directed or action-directing— may we speak of the simple, first stage of consciousness, when the state of brain is barely distinguishable from unconsciousness, as in a predawn, indefinite stage.

The short review of the dynamic of reasoning presented in Part III considered only the obvious relationships between the development of thought and its principal promoter, will. A precise and unitary definition of will comprising its essence and its mechanisms will also not be attempted here. However, we shall see next that will cannot be considered simply as a function of mind but as something more intimately connected with mind's basic substance (a still open question). We must recognize also that the purely metaphysical aspects of will are indissolubly bound to the operational ones through which mind interacts with physical reality. By this, will is deeply entrenched in almost every neurophysiological function, has a counterpart in nervous activities, and has its own share of neural circuitry.

Repeating here the arbitrary, partial, and operational definition given in Sect. 5.4, taking into account what we may call the heuristic, phenomenological procedure of will, one first finds the intention, which may or not be externalized in behavior. As an intention, will appears at every point of a line of reasoning, that is, it is integral to mind's eidetic activity, not simply its product —although mind appears to be its origin— and must have in itself a rationality as basis of its own nature. Considering will as a rational structure enables us to examine the possible logical and neural procedures eventually resulting in that intention. In Parts I and II we maintained that at every step in the reasoning procedure mind must make a choice of the succeeding step to be undertaken on the basis of a comparison between alternatives. This procedure was held to be fundamental as it represents an operational principle, realized in different ways in almost all brain functions (see also *PTN*), but no discussion of the underlying criteria was undertaken. Comparison, as such and independently of its specific objects, leads to a judgement, although not necessarily a conscious one. In many brain functions involving automatisms a judgement may also remain at subconscious levels and lead to an automatic expression of some behaviors, in such low-level functions as simple as opening a path excluding others (facilitation, *Bahnung*, inhibition, occlusion, etc.). The deciding factor, which may be the actual criterion at the basis of comparison, can be represented by the "weight" of each possible alternative —better, the difference (or the ratio) between their weights. If one accepts that this simple heuristic procedure (i.e., the weight criterion) is generally at the basis of each comparison and choice process, one could suppose very simply that the so-called "voluntary" choice implies consciousness only as a surplus, i.e., the conscious acceptance of the result of comparison, which in the extreme case could amount to nothing more than performing the judgement in the state of awareness or the recognition of what the comparison has produced.

All this leads to the conclusion that the fundamental structural organization constituting the operational substrate of will-intention can also be accessed beneath the level of consciousness, giving rise to an evident absurdity: an unconscious will. However, to conclude that will is operative, the role of awareness can certainly also not be restricted to simple recognition or acceptance of the results of a comparison. In fact, if one accepts that the criterion of comparison rests on the weights of the alternatives, one must also take into account how the weights are determined, and in this process a multitude of factors may have to be accounted for. The weight of abstract objects, each comprising a number of elements, is made up of an equal number of elementary weights, and each of these may be partly predetermined (i.e., there is perhaps also an inherited, genetically and specifically determined component); in large part, however, the weight results from factors related to general education and specific training, to knowledge, and to frequency of its use in eidetic matters. Moreover, noneidetic elements, such as emotionality, may also be present. This means that the whole preceding history of mind contributes to the formation of the voluntary decision, with the goals to be attained imprinting characteristics peculiar to each individual. Operationally, the acceptance of the goal itself may result only from the comparison between goals, that is, again from their associated weights, to which the preceding mind's history contributes to the largest part. Observation shows how the whole procedure evolves in time as a "refinement" (maturation?) of initial goals toward the definition of some final goal, an endpoint giving meaning to all mind's endeavors. From what has preceded, awareness is instrumental in determining will, as the latter is commonly understood. However, this raises the need to consider another aspect of will specifically connected to the nature of awareness. We may understand will not simply as conscious recognition of the results of a comparison, and as such appropriated by the conscious self, but rather as the expression of a force. This sense of will may easily refer to the less immediately evident meaning of awareness, that of guidance, implying that the latter has inherently the nature of a force directing mind's operations toward some (in mind) prefigured goal. Here, the concept of force may be applied in its proper sense, as in classic dynamics (for instance, Newtonian motion), where force is defined by mass and velocity. Likewise, in eidetic activity the weights of different objects also have an associated intrinsic momentum, the intrinsic significance of each weight in relation to the situation and to the preceding history. The weight is in itself the inertial mass (reached in different ways —inheritance, training, etc.); the velocity (and the acceleration) is produced by the activation of the mass in view of the goal. We can advance the assumption that what we call activation of the mass is the actual will.

How is the mass activated; how is the activation operated; or more precisely, in what form do we figure the generating mechanism of that motion-activator constituting in our hypothesis the essence of will? There may yet be no reason for abandoning the hypothesis that in the case of will conceived as a force the basic comparative procedure also provides the mechanism of decision. In fact, we already have suggested the existence of an eidetic force, created by a gradient of energy inherent in the eidetic contents of two abstract objects (identifiable in the intrinsic contents of the two reasonings). As before, the conscious recognition of the eidetic content and therefore of the gradient and its direction is sufficient to determine the "acceleration" of the weights relating to one of the objects, hence to determine will and its "direction."

10.3.1 Will, the *Ich*, and its Structure

Let us examine more closely the latter hypothesis. As was said above, the mechanism of comparison and the subsequent judgement need not involve awareness; as a comparison, it is a basic mechanism common to most cerebral operations, including those performed at an unconscious level. Consequently a high degree of automatism may generally be inherent in comparison and judgement. However, when eidetic matter is concerned, awareness cannot be dispensed with since, according to our hypothesis, the simplest of procedures that can be envisaged for a truly voluntary choice requires the conscious recognition of the results of comparison, that is, the judgement. Nevertheless, even in the presence of awareness a simple act of recognition would mar the concept of will as an essentially free act of decision as we generally conceive it because the decision would be left exclusively to the inescapable logic of the reasoning. In this event, the available degrees of freedom are extremely reduced since the logic inherent in the quantitative relations among the weights of all the elements still represents the most relevant factor, and mind thus behaves almost passively. Therefore, if we assume the existence of a will that we may call "real" and independent of conditioning not only from without but principally from reasoning itself, we must recognize that "real will" equals what we generally call "free will." Before entering the more metaphysical aspects of this question, we must attempt to identify, if not in the usual neurophysiological terms at least in psychological ones, the entity having the capacity of exerting free will.

No degree of predetermination inherent in the weights of the elements at the basis of will and no energy gradient intrinsic in the reasonings can withstand the "negation power" of the generally held concept of free will, which can oppose and terminate whatever motion may arise in conscious mind as a result of weights and gradients (see Sect. 5.4). The capacity of negation may stem only from a source deeply rooted in an "entity" itself having the ability also to oppose logic and emotionality, to negate rationality and sensation, and to withstand the strongest drives of the inner and outer worlds. One entity alone possesses these properties: the *Ich*, the ego, the root of the self. We only have to assume that this entity negates because the negation of the "other-than-itself" is an expression of its own determination, of affirming its own existence and right to freedom from all possible determinants and conditionings. The basic core of this, however, lies in the freedom of the *Ich* also not to negate, even if nonnegation may not comply with some of the aspects of its life consequent to its own determination. The single inescapable constraint for the *Ich* is the impossibility of autonegation, since every attempt at self-negation would still represent an affirmation of its own existence, its own freedom, and once more results in its own determination.

By interpreting will as fundamentally an expression of the *Ich*, we reach an entity that cannot be observed from without but only from within. No external observer can gain direct knowledge of the *Ich* and its properties; it can be observed only as an aspect of self-awareness. The knowledge that self-awareness can gain of the *Ich* does not derive from comparison of the properties of the *Ich* and the non-*Ich*. If this were so, it would be the same as assuming that the *Ich*, as the result of a comparison, could not be held as the ultimate term of reference since, being representative of a relative judgement, it would likely change with the changing properties of the myriad non-*Ichs* taken individually for comparison. Therefore, self-awareness can gain knowledge of the *Ich* only as the fundamental substance of the latter's nature, i.e., an extreme "generalization" beyond any specification; in this way, and only in this way, can the *Ich* be a reference term for every non-*Ich*. Of course, this "hard" *Ich* maintains continuous relationships with the great variety of non-*Ichs* of the world, in the form of negation or nonnegation. In this case this can take the form only of nonacceptance or acceptance, obviously not of the essence, of the "hard" non-*Ich* that would not be ultimately knowable, but of the explorable part of the non-*Ich* nature, i.e., its free-will. This is the only means of expression available, the sole "language" through which the *Ich* communicates with the myriad non-*Ichs*.

Negation and nonnegation are the external signs of will, as the force that does not result from a comparison, i.e., of will-guidance. Accordingly, the *Ich* as source of will-guidance cannot be a fleeting entity, as it would be if it derived its properties from its relationships with the innumerable animate and inanimate essences of the surrounding world. Thus, the *Ich* is not a mere essence, the set of properties pertaining to an *esse*; is *the esse* itself of the individual, as stated above, the extreme generalization beyond any specification. The *Ich* is the "individual act," i.e., the *esse* which has become "Dasein" and to which the essence gives the specification of the individual, that ensemble of properties making each individual *esse* what it is, i.e., different from others. Thus, the "hard," or "integral" *Ich* has the substance of the *esse* and is the hard substance to which the essence —as ensemble of specific properties— refers. However, since the essence of the individual comprises the material as well as the abstract properties constituting mind-brain, the *Ich-esse* to which the essence refers must comprise both materiality and abstraction, i.e., its structure must partake of both natures as mind-brain does.

These conclusions suggest an answer, originating from physiological premises, to a question which on purely philosophycal terms is still open to endless discussions and interpretations. On the one hand, the immediate *Ich* (the "hard" *Ich*), a substance independent of the surrounding objects and their specifications, an act, a primary substance which imposes itself (is will in act) by virtue of the mere force of its existence, but an *esse* necessarily endowed with determining characteristics, i.e., with own essence. On the other hand, the *Ich* existing in the world in relation to other objects (non-*Ichs*), a "mediated" *Ich* becoming, through its own essence in relation with other essences, a relation itself, no longer "hard" *Ich*, a mere product of mind's experience, a nonprimary, non-independent, fleeting entity, as fleeting are the experiences concerning the essences.

10.4 Conscience

Whenever aware mind has the guidance properties typical of will and chooses a line of thought, it cannot do so at random but performs a selection on the basis of some preceding eidetic process and toward some goal. This is to say that aware mind in some way must contain in itself, or at least must have access to, the goal. For this there must be present in mind (at least in the form of simple outlines) not only the problem but also the possible developments, thus enabling it to select between alternate solutions. Furthermore, aware mind must also have access to an overview (even if incomplete) of the state of the phase space in terms of available ideas in order directly to promote an exchange with phase space, seeking the correct ideas to be implemented in brain space-time as well as looking for already implemented ideas in the brain circuits. In this way the state of awareness acquires for mind an aspect of self-examination. However, according to our theory, this is a function that an aware mind cannot accomplish alone. If it is possible at all, this can occur only through the close relationship between awareness and consciousness (these are, according to our definition, two stages of the same state of mind in brain space-time, or more simply two properties of mind) and, in turn, between consciousness and conscience. In fact, as a first stage consciousness may exhibit its peculiar property of acting as a bridge between brain space-time and the phase space, i.e., of allowing mind to reach its own essential abstract roots. Certainly, the procedure cannot be in any way perceived, as these operations are carried on at the first stage, i.e., simple consciousness; this would exclude awareness, but the results would be revealed by the "appearance" of the ideas in mind when awareness is resumed. Mind would continually "oscillate" between these two stages; once again we propose that mind can be in two states simultaneously (better, in two stages of the same state).

The existence in conscience of concept-ideas together with all their cross-relationships means that potentially inherent in it is the network of all possible paths that thought may follow, and therefore also comprised in it are the possible goals that may be pursued. For these reasons, conscience can no longer be considered simply as the catalogue of ideas or even as the abstract compendium of the individual's knowledge. Knowledge of the possible paths connecting ideas entails presence of the logic inherent in the paths, and this must provide at the same time the standards of comparison for the operative side, i.e., for the eidetic processes in brain space-time. Thus, conscience has all the properties necessary for making it the first judge and guide of thoughts and acts of man. But, as just stated, since it is a phase space and not a state of brain or a function, no process can take place in it, and therefore all its properties remain unnoticed until they may become apparent in awareness (i.e., in directing, or directed, consciousness) through which the properties of conscience interact with the space-time of brain, without actually being linked directly to it. Although mediately, from conscience originates all information necessary to aware mind for directing the brain's acts, i.e., for expressing will —an inescapable conclusion once we accept the properties of the eidetic phase space. Thus, the nature of conscience is that of a "being in potency" encompassing all the potencies of mind and of which in physical space-time life we can see "in act" only a very small part at a time (in the state of awareness) and never all of its innumerable possible facets. If we accept that conscience phase space constitutes a region outside physical space and not temporally connected with it, a region that can be spanned instantaneously in all its aspects, then knowledge is hidden in the intricacy of the relationships linking the concept-ideas to one another, and remains for large part unknown and unnoticed for most of mind's life.

Thus, conscience is not the simple compendium of that which has been introduced into it in terms of natural, constructed, and abstracted ideas. By far the largest part of the eidetic wealth is already implicit in the logical nature of every concept-idea (the field properties, the innumerable essences obtainable by abstraction procedures, the ideas still hidden in the maze of intrinsic relationships, and those forcedly brought to being by logically filling the gaps in reasoning, etc.). Accordingly, we must recognize that conscience has in itself everything that can be desired in terms of intellectual function, but left alone still is not an intellect in act. It would therefore be altogether preposterous to try to envisage the conscience just outlined as existing independent of its own mind and its properties of consciousness and awareness. Conscience, in this view, is a power, albeit in potency, but always a power with a primary source retraceable in matter; therefore, asserting that conscience could exist without mind-brain would be like accepting that gravitation is possible without the source of its energy, i.e., without matter (see below).

10.4.1 Constructing Conscience

Upon these arguments we should reconsider the way in which the wealth of concept-ideas constituting the phase space takes shape (see also Sect. 9.5.1). By genetic determination one can inherit not the ideas themselves but an anatomical structure suited for embodying and developing ideas. The brain, after a period of structural settlement —what we have called anatomical and chemical maturation, see *PTN*— can implement ideas because it possesses the physical structure of the circuits (inputs and feedbacks). However, functional (and structural) maturation is that which is actually responsible for the formation of ideas in the brain because through exercise (practice) and learning the maturation of the numerous differently organized networks of synaptic arrangements in the circuits of neurons is promoted. Due to the slowness of the procedures only awareness can be involved in the formation of ideas, while the transposition of these into phase space is a function of consciousness, which alone can provide the instantaneousness of operation necessary to deal with phase space, introducing in it the nonorganized bulk of ideas (better, organized according to the brain space-time causal logic but nonorganized according to the timeless logic of conscience phase space). Nevertheless, the picture would not be clear enough without pointing out that phase space conscience represents our whole knowledge, paistakingly built, at first by acquiring concepts and the most primitive ideas about the surrounding world (natural ideas) and most necessary for surviving in it and only later by constructing ideas for a more sophisticated and differentiated social life as well as for satisfying the needs of intellect. However, the elements for assembling constructed ideas can be obtained only from the phase space, in which the items are ordered according to its logic. Thus, the possibilities of using the phase space (of realizing conscience "in act") necessarily undergoes a parallel development with its progressive construction

Recalling Sect. 10.1.1, the ontogeny of conscience consists of the appearance of concepts representing potential blocks of ideas; their becoming linked by the logic of the phase space (that logic inherent in the timeless relationships among the concept-ideas) marks the culmination of this ontogenesis. The fundamental property of each block is to comprise the extremely articulated complex of relationships among a large number of ideas at every possible level of abstraction centered about a core of contents by their intrinsic coherence. The system of relationships constituting the logic that links the ideas of the concept is recognized by mind as forming a unity in itself. However, on the one hand, the system of relationships cannot be recognized and understood in its entirety due to the intrinsic complexity of the objects it represents, while, on the other, it imposes itself on mind, and mind "accepts" it but only for the part that can be understood, with characteristic immediacy as one not requiring demonstration or explanation. The concept is thus a self-explanatory abstraction, the only one with a reality, but never sufficiently "clothed" with ideas. The concept is taken as an apodictic, undemonstrable truth, or at least as a truth of which one may give only partial and tautological demonstration (see Hegel, *Grundlinien der Philosophie des Rechts*, "Introduction") and a truth with somewhat different meaning according to the different ideas with which each mind clothes its own concepts. The point I want to stress here is that although all the ideas can also be thought of as having a distant origin in nature, the concept cannot be created in the aware mind, or given, or suggested to it by nature; rather, it first appears in the phase space and results from a system of relationships between the constituting element, brought to fruition by a suitable stimulus. Nature per se represents only one of the sources of such stimuli.

The preceding paragraphs emphasize the instrumental role of the brain circuits in primarily constructing the phase space as they supply the phase space with the first elements from which a simple concept may arise once their noncausal, timeless relationships are ordered by the logic of intrinsic coherence of contents. In this sense it is correct to say that ideas are not present at birth, not even in potency. Mind, at birth, is a tabula rasa, a blank sheet onto which nothing can be written until the necessary maturation of structure and of organization have prepared a number of eidetic circuits. Nevertheless, it is also logical to presume that there are some predetermined eidetic circuits at birth, precisely those related to fundamental natural ideas assisting the individual's life in its first confrontation with the outside world, for instance, those concerned with receiving the "imprinting" of associations between stimuli and fundamental patterns of behavior. For this reason, we should see this preformed substrate not necessarily as involving eidetic processes (a kind of a reflex arc arrangement would be suited for most purposes, and much of the imprinting characteristics seem to suggest this mechanism). At any rate, the inherited component should be basically essential for animals much more than for man. In fact, newborn animals are generally very early left alone to confront their surroundings, so that a built-in function appears indispensable. As has been repeatedly asserted, man's slow maturation of body and mind shows the necessity of an extremely long training in contacting the world as well as in constructing and using his eidetic phase space, i.e., it shows a slowly growing conscience function. Nonetheless, to a larger or lesser degree, even if still not clearly defined, the eidetic circuits must be outlined rather early in life. All their synaptic structures are certainly not yet completed at birth but are progressively added in the correct patterns along with the structural and chemical maturation promoted by the same training. Therefore it seems more than justified to say that the eidetic function is in potency in the brain at a very early age. What gradually contributes to differentiating various brains is the differing use made of this potentiality. Thus, once more, the emphasis is placed on the mechanisms which realize the potency, bringing the ideas into "being" so that they can be used or experienced in the brain circuits.

10.4.2 Hypotheses About Phase Space

Before further discussing the mechanisms of "realizing" conscience, we should deal further with the alternative hypothesis about the nature of phase space. Once we are forced to accept that eidetic structures may be genetically determined and are therefore present at birth, we can reconsider the question of the reciprocal interaction between conscience phase space and brain space-time. Essentially, there is no need to posit an abstract (conscience) phase space as an object different from brain space-time, with all the undeniable difficulties in understanding how conscience communicates with this. Since the eidetic circuits are already there, the alternative hypothesis postulates that experiencing the ideas (the assembling, analysis, ordering, etc.) relates exclusively to space-time itself; the abstract functions in this case should be autonomously implemented in implicit form, without a conscience "feeding" them from the phase space into the brain matter. New ideas would be experienced by testing them directly in the matured cerebral circuits, that is, they would be newly assembled from elements or from ideas already existing in act or in potency in one or the other of the many circuits available. This interpretation sees the process of idea formation differently, in a way more pleasing the empirical, positivistic scientist. In fact, if one accepts the presence, at some time, of all the circuits and inputs (either activated or not), new ideas would not need to be created, for they would simply be found, recognized, or at most assembled, among the innumerable possible combinations presented by the implementing circuits. Collecting the circuits could be operated on the basis of their inherent congruence and intrinsic coherence with the line of thought followed up to that point. This hypothesis, that alternative to the hypothesis of time-unrelated conscience phase space, sees the brain space-time itself as put into act by a different type of conscience; every eidetic operation would begin in space-time, in the state of awareness, and the idea phase space would be constituted by a truly physical space —the ensemble of brain circuits thus containing all the ideas, either "in potency" or "in act." In other words, an abstract phase space unrelated to time would no longer be needed. The phase space would have a physical consistency; it would have a beginning, a slow development and a continuous time-bound transformation into itself, i.e., it would exhibit a time-dependence as every other material, brain-bound function. As far as the ideas themselves are concerned, they still would be "immanent" but in the restricted sense of being in potency in brain space-time only.

Such an interpretation of the eidetic procedures concerning the source and formation of ideas would no doubt appeal to most neuroscientists, at least to those accustomed to dealing exclusively with the "physical" brain as the only one permitting verifiable —physically measurable— facts. It is surely sound to reject hypotheses that cannot be submitted to experimental test (as the real existence of an actual, not purely mathematical, and yet inherently immaterial phase space of the ideas). Unfortunately, at least as far as the brain is concerned, there also are objects that are not measurable in physical terms or as yet cannot be submitted to physical measurement with universally accepted standard units of measurement. However, a physical phase space, i.e., a space with a function comparable to the mathematical concept of abstract phase space, although relating strictly to physical space-time, prevents the brain from constructing nonnatural ideas by itself and harboring the concepts. In fact, we said that constructed ideas are created with the purpose of filling logical gaps in reasoning. A constructed idea is not concerned with natural elements obtained from outside but with element-ideas of different sources at a higher level of abstraction. Preknowledge of the outline of the goal to be reached is thus required for overcoming the logical gap; therefore —and here is the point— preknowledge is necessary also of the idea's content. This explains why that preknowledge can be identified, at least in large part, only with the concept that mind proceeds to clothe in ideas (see Sects 8.1 and 8.4) —a concept that mind cannot construct by itself but which "imposes" itself on mind in an immediately presenting and graspable way (see the Sect. 10.1.1). If the formation of concepts were a process exclusively of brain matter, one would encounter no difficulty in clearly and exhaustively describing every concept, since all element-ideas would be immanent in mind and no gap could exist in the logic of reasoning because all the needed elements would be present and readily available in space-time language. Moreover, for these reasons, interindividual differences regarding the concepts' essence would be reduced almost to nil, that is, almost everyone in the world would agree on virtually each issue which presently gives rise to conflict and debate. Unfortunately, however, such a clear and exhaustive perceptibility of concepts is given to no mind, and this renders the hypothesis of nonphysical phase space necessary to explain the existence of concepts and the wide differences in judgements about the same objects; the construction of new ideas for filling in logical gaps in reasoning appears to require the properties peculiar to conscience phase space.

There still is the question of the speed with which the trained eidetic system responds to the demands of reasoning. We have repeatedly said that there is an instantaneousness in the appearance of ideas and in realizing affinities among them, as if mind already knew them. It is not easy to understand the mechanisms of the immediacy of the eidetic processes within the framework of a physical phase space. Since this is necessarily confined in brain space-time, the related eidetic processes can occur only in the real time of brain circuitry. Therefore, one must account not only for the conduction time along nerve fibers but also for synaptic delays and particularly for the decision times after the innumerable comparisons needed to detect the congruence of elements and the coherence among ideas. Thus, if one accepts that "flashing" speed (suddenness) is a peculiar property of many (though certainly not of all) eidetic procedures, this can hardly be reconciled with a function which at every step of its development relies only on nerve cells and fibers. It can be observed that a relativistic deformation of space-time when mind is strongly engaged brings about a considerable slowing of personal time (see earlier sections, and *PTN*). The arguments concerning relativistic phenomena might account in extreme cases for such an extreme slowing of personal time, such that one supposes the existence of suddenness in eidetic processes. However, as we have seen, the logic of relativistic transformation of space-time during mental engagement accounts for a slowing of the system's time —the *Eigenzeit*— which can be perceived only with reference to an external observer, i.e., an external clock. On the other hand, the time relationships within the system suffer no strains because every part of the eidetic structure behaves according to the structure's own time, i.e., it covaries with it. Moreover, all our previous arguments have brought us to the conclusion that the eidetic system defines the metrics of brain matter space-time, and not vice versa (see Sects. 2.3, 4.1), and there is therefore no independent observer within the brain that could perceive that suddenness. Finally, the demonstrated personal time deviations are never so large as to be misjudged for "no-time" suddenness (although surely what we call personal time may constitute simply a scaled, up or down, representative of the true *Eigenzeit* of the eidetic system.

D. Rakovic has approached the problem from a very different point of view and developed a novel hypothesis. This is based on the high magnetic conductance of a suitably structured (very low dielectric) medium, through which the electromagnetic component of ultralow frequency brain waves propagate with negligible loss of energy. In this way a quasisimultaneous presence of "objects" is assured throughout the brain (this might represent an equivalent, inherent in the still imperfectly known physical properties of the neural tissue, of the transformation that makes the points of the region outside the cones simultaneous with the "present" at the cones' apex). According to Rakovic, this is an electromagnetic-based hypothesis about the nature of consciousness which also accounts for relativistic velocities in a number of different brain conditions and the possibility of perceiving time deformation by an observer inside the brain.

The hypothesis of an eidetic procedure entrusted exclusively to cerebral space-time, i.e., without an abstract phase space in the background —even if it apparently avoids the necessity of postulating the "real" existence of purely abstract objects— does not actually eliminate it, nor does it eliminate the abstraction which produces the invariant abstract objects. In fact, abstraction represents a very special transformation concerning a real object in brain space-time, i.e., an idea implemented in the neural circuits. However, this special transformation leaves unaltered the object, while it produces a new object; to do so it requires an eidetic operator having a real and an abstract side, the invariance being a property of the abstract component (see Sects. 8.7, 8.8). Therefore, if the invariance of ideas is accepted, it must be associated with the abstract side of the operator and with the abstract side of the idea, by which we are led to consider as "real," in the sense of "really existing," the sole abstraction of operator and idea, whereas the physical, matter-based operator and idea are bound to vanish at any time, for instance, as the circuit is opened. Consequently the eventual discovery of the essential nature of the object in its purest abstraction could not be accomplished in cerebral space-time alone. All ideas concerned with justice, society, good, etc., could not be formulated if the corresponding originating concepts —essentially and exclusively abstract objects— of justice, society, good, etc, did not really exist, albeit abstractly. In fact, the latter, before being clothed in the ideas which represent them in space-time must exist for every individual in a personal phase space consisting of the complete set of relationships among ideas and other concepts, including those that the brain's space-time eidetic system is unable to understand exhaustively. The common experience is that, no matter how much the eidetic function implemented in brain circuits tries to express it precisely, the concept continually eludes successive redefinitions, refinements, recalling of other concepts and ideas, analogies, etc. In our opinion, this can only mean that the concept, although really existing as an abstraction, is structured according a logic different from that of physical brain circuits and is hindered in its becoming real —and only in this— as it cannot really exist in space-time as a concept as such, i.e., as a timeless object with the timeless structure of noncausally ordered phase space conscience but only as a set of ideas developed from it and only imperfectly defining it, i.e., as objects causally connected in space-time. What makes imperfect the transformation from abstract concept to actual ideas can only be the imperfect transformation from the logic of an internal language in which the concept is shrouded (if we may say so) to the logic of the external language of our thoughts, actually a different logic, the logic of brain matter space *and* time, the causally connecting logic, the only one that can physically express the abstract objects and their relationships.

10.5 Will Realizes the Eidetic Function in Potency

If the hypothesis of the exclusive existence of a purely physical conscience phase space is discarded on the basis of the preceding arguments and the objections concerning time, we are left with the former hypothesis of ideas immanent in a region not physically connected with brain space-time from which they may enter the brain circuits. In the latter hypothesis, ideas are immanent in a time-independent way, either in potency, if the circuits are open, or in act, although not used in eidetic procedures, if the circuits are closed but not under the lens of awareness, as though dormant in the brain matter until some agent intervenes. The point is that an agent of some sort is also required by the hypothesis of the exclusively material phase space in order to account, on the one hand, for the nonchaotic functioning of the eidetic system at every step, particularly when initiating reasoning, and on the other, for an independent eidetic operation, i.e., for the not driven, autonomous functioning of mind (unless one wanted to deny the latter possibility). Concerning the nature of that agent we have hypothesized that it could be identified with will. Consciousness and awareness together with conscience, as constituting a functional unity having both abstract and physical structures, are concerned with the voluntary act of selecting lines of thought to be implemented in the brain circuits, and developed in time.

Let us return now to the problem of the relationships between will and the eidetic function. We have focused the problem: from how conscience, as abstract phase space of ideas, can influence brain space-time, we have shifted the enquiry to how the will of transformation from abstract phase space to physical space time can arise —as yet, assumed to be a directing function with hardly more than a label.

Will represents the agent necessary to set in motion the procedures; without it, consciousness, awareness, and thought development could have no place in brain space-time (unless one could conceive an idling and spontaneously moving, purposeless eidetic, i.e., ideas that keep running by themselves). While from a purely technical viewpoint a spontaneous motion could well occur (i.e., procedures set in motion by the random closure of circuits through the innumerable inputs to the system), such an unorganized activation of eidetic procedures does not seem likely to be able alone to produce any ordered sequence of ideas and certainly does not represent a fast or a successfully and reliably working mechanism. The same abundance of inputs to every circuit, partly shared with other circuits, represents by itself a hindrance to the logical ordering of any particular line of ideas because every eidetic line rests on the selection of sets and sequences of ideas excluding all others, a feat presupposing, as already seen, a preknowledge of some sort about goals that the chaotic activation of the inputs would be unable to ensure to any extent. One could argue that a naturally patterned activation of inputs could easily be determined by the natural laws superseding the outside events and thereby the conditions of the surrounding situation. Truly, this represents an everyday occurrence, but it cannot account for ideas or lines of thought unrelated to situations outside the brain. The eidetic system enjoys an autonomy of beginning and of development, independence from what may drive it from without, all signs of a real freedom, and thus requiring the existence of a real free will (see Sects. 10.3, 10.3.1). Even the philosophers of ancient and medieval times affirmed that will resides in rationality; the principle that we advance now is not that will "needs" rationality, or that rationality (i.e., the principles of organized thought) can exist independently of will, but that will —as it expresses itself in eidetic function— and rationality are the same thing.

Returning to the transposition of abstract phase space into the physically realized one, will can be seen as naturally joined to "act;" therefore it presupposes consciousness, i.e., realized conscience. On the other hand, consciousness ought to be directed, or to direct attention, toward the will-indicated specific region of ideas; hence simple consciousness alone is not sufficient fully to realize will. The necessary condition here is awareness, since awareness alone is concerned with particular eidetic regions of knowledge and action. Considering the process at the level of brain matter, awareness is necessarily and specifically directed (like a beacon light) toward the circuit implementing the idea that we want, or must, be aware of. Thus, awareness and realized will, unless one wishes to keep them separate for other purposes, must be considered as merging into a single neurological object, a function-state. Moreover, it is obvious that we must recognize in these will-awareness realized circuits the primary origin of whatever decision or action results from the ideas becoming real (themselves the expression of will-awareness), such as the development of a line of thought. The unavoidable conclusion is that will involves the whole ensemble of eidetic circuits and is in potency in them, as are all the ideas that can be embodied. Furthermore, together with the ideas the lines of thought are present in potency that can be organized on the basis of the available ideas, and that will-awareness can eventually put into act. At this point, one may wonder how it is possible to discriminate between the state of awareness of the brain cells —thus of the eidetic circuits— and will and ideas embodied in them. At the level of nerve cells, the aware state, the will function, and the objects originating in that function —the embodied ideas— merge into the same organized cell activity, which thus represents them all at once.

We should point out that the being in potency of ideas in conscience phase space, as noted above, and the being in potency of will-awareness in brain space-time have altogether different bearing as far as eidetic activity is concerned. The difference comes from the constitution of the two spaces. In conscience, ideas are elements of superior, organized objects —the concepts— and these are elements of global, integrated knowledge, which can be reached fully only by comprising in it all the relationships among the concept-ideas available in phase space (we should recall that conscience literally means simultaneous and integrated knowledge). The crucial point of difference between conscience and its realized counterpart in space-time, i.e., consciousness in its stage of will-awareness, lies in the fact that in brain space-time the sole knowledge that we can be aware of consists of the ideas embodied in the circuits, the only one that we can realize voluntarily, while a large part of knowledge remains unnoticed in conscience and is unusable by will-awareness (since it cannot be embodied in matter, consisting of relationships that mind is as yet incapable of understanding). This does not exclude that through painstaking training and learning one may arrive at a clear understanding of other and more complex patterns of relationships between ideas of phase space, to be realized further in the space-time of will-awareness, thus expanding the knowledge that can be utilized. However, we shall humbly recognize with *Wagner*:

Ach Gott! Die Kunst ist lang,
Und kurz ist unser Leben./...
Und ehe man den halben Weg erreicht,
Muß wohl ein armer Teufel sterben.
(Goethe, *Faust*, v.558-59, 564-65)

10.5.1 Heuristics of Conscience Realization

If will-awareness can at any time initiate an altogether different and new line of reasoning, we must consider the mechanisms through which conscience enters into act, first becoming consciousness, then eventually awareness and realized will; as a final result, certain circuits are selected and others excluded. The function that we called above "lens of consciousness" comprises these operations. It is clear by now that consciousness and will-awareness represent two stages of the same state of brain, whereas conscience as phase space is not part of space-time and is not considered a brain state whatever. Furthermore, the definition of conscience as a phase space has been useful only for comprising a number of properties seeming to befit a phase space. However, as it appeared in the course of the discussion, while conscience as phase space maintains the properties of the corresponding mathematical "object," conscience in its proper nature is a metaphysical concept. Nevertheless, if we want conscience, consciousness, awareness, and realized will to partake of, or be viewed as the same reference framework, a unique line of directing action must be envisaged from conscience to act, and a possibly unique agent must take care of the transition from abstract phase space to space-time, as well as throughout the latter's domain to provide for the realization of will in outwardly directed actions.

The properties of conscience as phase space of ideas coincide, of course, with the metaphysical concept of conscience as knowledge. The latter also has a material support in the brain, as the ideas forming the basis of knowledge are simultaneously experienced by will-awareness in the space-time of the brain circuits, but we have said conscience knowledge is larger, more comprehensive, and more integrated than will-awareness knowledge (since the latter cannot include all the intrinsic relationships linking the concept-ideas). Nonetheless, the two types of knowledge must still have a fairly sized, common content. In fact, until the abstract relationships among ultimate essences and concepts are disregarded, we can take will-awareness knowledge in space-time to be a fairly good representative of conscience knowledge. This is of great help when dealing with the several properties of conscience, because the existence in conscience phase space of relationships (among ideas) not yet understood by mind constitutes a fairly sound assumption as theoretically necessary but still not a demonstrable fact to the eyes of the positivistic empirical scientist. Monitoring the "life" of conscience is prevented by the contradiction in terms with the definition of phase space (in it everything is always present and nothing can be going on, not even a scanning from the outside). Conscience, as we envisage it, is global, integrated knowledge of its components, the concept-ideas, in their ensemble constituting the "maximum science" connected with them. The possibility of directly monitoring conscience would imply that optimum knowledge could be directly reached without any further scientific investigation. On the other hand, if every system of relationships existing in conscience were readily understood by mind in the state of awareness, conscience itself would immediately cease to exist necessarily, and with it also the concepts (real, though abstract, entities), which we said may exist only in abstract phase space. Conscience would be only an unnecessary duplicate of that which awareness realizes in brain space-time, for all ideas, once understood and primarily, directly embodied in brain matter, could exist there, including all their intrinsic reciprocal relationships that mind is able to understand.

The task of gaining knowledge of conscience through an inner exploration of it now appears inconceivable. However, we shall seek a general scheme of the possible operational relationships between conscience and brain space-time. Notwithstanding all our arguments up to now, our notion of the abstract structure of conscience is still not much more than a confused and equivocal mataphysical construction, looming outside space-time, hence, not embodied in brain circuits and yet "delivering" to these a part of its own content —we can now say part of itself and for this very reason making itself real. To the latter purpose, the postulated bridging properties of the simple, first stage of consciousness would allow transition from metaphysical no-time to physical space-time. At first glance, this rough outline seems to conceal contradictions, but I hope to demonstrate that these are more apparent than real.

Let us consider first a most natural hypothesis based on the common image of conscience. That which could make conscience real —constituting at the same time the basis of will— could be a choice made by conscience itself. The latter, by the fact of dwelling upon (or directing itself to) a given set of ideas makes them real, i.e., it would allow the cerebral structures at the first stage of consciousness automatically to receive them into a phase space embodied in the space-time of brain, then rising up to the stage of awareness. In this way, the transfer of the ideas into space-time is accomplished. It should be stressed once more that while consciousness and awareness have quite different properties, they are nonetheless two aspects of the same object, and therefore that whenever awareness is aroused it does not necessarily exclude the simultaneous existence of the bridging property of consciousness at its first stage; this allows the continual transfer from phase space to space-time. In fact, we can experience in ourselves that awareness always rests on the substrate of simple consciousness (of the self existing together with the world around it). However, the contradiction *in re* with the substance of the definition previously given of conscience and its inherent properties, is immediately evident. When talking of choice on the part of conscience, we mean an act, an event, some change or directed motion by definition relating only to space-time, therefore to a conscience already realized in a specific area, and thus only to consciousness-awareness.

Will, moreover, is necessarily implied in every eidetic choice, and in this way it is present in conscience not only in potency but also in act. In turn, will in act presupposes that comparisons have been made between alternative solutions, an act that could not be carried out in conscience. All of these features contrast sharply with the postulated phase space, timeless characteristics of conscience; therefore, if the latter are still taken to characterize it, the heuristic procedures of communication from conscience to consciousness and from consciousness to awareness must be envisaged in another way. The main point of contrast lies in the voluntarily operated choice of ideas in conscience and the voluntary transfer of them into consciousness. The fact that the parameter of time does not relate to conscience phase space, and this bars altogether any relationship with events in brain space-time, does not prevent conscience from being logically ordered. It is actually all too obvious that in phase space the ideas belonging to the same region, or to fields of the same nature, may form subspaces available as blocks. Consequently, we picture the phase space as consisting of a large number of such blocks, simultaneously present to, and therefore utilizable by, consciousness-awareness. Thus, sets of concept-ideas are not actually selected (we shall not use this term any more) by conscience, but the conscience phase space is made up of blocks of inherently congruent concept-ideas —let us now say naturally ordered, namely according to their "natural" reciprocal relationships, i.e., to their intrinsic coherence. Thus, we may consider coherence as a force keeping the concept-ideas ordered in so many blocks having the potential for logical essences, although not all fully understandable. Through their essences, the block concept-ideas may exert a particular type of "static" influence, i.e., not involving motion or acts and particularly no time dependency, on corresponding coherent lines of thought implemented in brain circuits in the state of consciousness-awareness, a gravitational attraction-like influence between the constituent elements of the two spaces, which represents the link between phase space and space-time. A transformation, as we said earlier, necessary to account for the communication between the region outside the cones and space-time at the cones' apex may take a form suggested by the gravitational attraction-like force field present between the two spaces.

This hypothesis is not a substantial change from that which has already been said in the present and in the preceding sections, but suffices to simplify the logic at the basis of the hitherto unknown neurological mechanism initiating the eidetic procedures in consciousness awareness, and at the same time to understand how thought may proceed with greater speed in brain space time. In fact, the voluntary selection of a line of thought can be operated only by will-awareness (the lens of consciousness) among the different possibilities represented by the blocks of concept-ideas simultaneously present in conscience. A particular set may be selected by consciousness awareness on the basis of two factors: a) the intrinsic strength of that field of attraction created by the coherence among the conscience block-ideas and the lines of thought physically implemented in consciousness awareness; b) the needs of eidetic activity in consciousness awareness to match the engagement of the nervous system in the particular issue at that moment. According to the first factor, consciousness awareness "feels" the strength of the knowledge and the rationality inherent in individual lines of thought, and a conscience block of ideas is accordingly selected. The selected sets of ideas, in turn, are simultaneously and instantaneously transferred into space-time, endowing the thought processes of brain space-time with that quasi-flashing property otherwise so difficult to account for. To the second factor is connected the question of the neural operations underlying the choice of the blocks and their transposition from conscience to consciousness awareness. If we accept the hypothesis that the true voluntary choice is made in space-time, again we can only take the fundamental procedure of comparison as that allowing a judgement and then a true choice among many possible objects. Comparison is based on congruence-coherence which, when inherent in element ideas, allows quick recognition of sets (or subspaces) of them, with their reciprocal relationships now ordered, however, according to the sequential logic of space-time, i.e., in form of a reasoning sequentially —causality and temporally— developed. We should not forget that the actual choice stemming from comparison strongly depends on the "weights" that mind attributes to the different sets of concept-ideas, quite independently of the latter's intrinsic congruence-coherence, while conditioned, instead, by the preceding mind's history (education, knowledge) and by the strength of emotional drives. Returning to the principle that all the brain's operations are regulated by the same type of laws as are other physical events, we can say that the voluntary selection of lines of thought and actions may be largely almost automatic, determined by the algebraic summation of the weights connected to the various factors involved. It is important to note that this weight principle is far from representing a passive, mechanistic acceptance of the results of comparison: in fact, mind's full active and decisive participation consists in autonomously defining the "subjective" weight to be attributed to individual sets of concept-ideas and of drives in the state of awareness.

Mind's full freedom is guarded by the latter operation, and only the *Ich* can overwhelm it with its own negating willpower.

10.5.2 Dialectics of Conscience and Physical Phase Spaces

The preceding section suggests something different from the possibility of understanding *all* logical relationships among ideas. This possibility is not given to mind in consciousness awareness because its capabilities for understanding are limited by the finiteness of the available neuronal circuits and by the logic-penetrating power of the underlying comparative procedures. To understand new and more complex relationships, the logical links among ideas missing in space-time must be filled in at the anatomical and functional substrate appropriate for the two-way transition between conscience and consciousness awareness. The missing links can be seen as items in phase space without a "code name," and consciousness-awareness cannot refer to these missing links from conscience to space-time until the code name becomes embodied (this procedure we called above "constructing ideas;" see Sects. 8.1, 8.4). This figurative way of presenting the problem, borrowed by analogy from computer practice, actually conceals strong similarity of mechanisms, as we shall see.

We must not forget the operation of transferring the ideas from brain space-time to phase space. It is most obvious to advance the hypothesis of a symmetry, at least partial, in the proposed principle regulating the transition from phase space to space-time, i.e., a "gravitation-like attraction" between the ideas in phase space and the ideas experienced in space-time. We must therefore better define the beginning of phase space in the individual brain. If the transition requires a constant attractive force of some sort, this must be present from the very beginning, when conscience is still empty; this is clearly preposterous. However, we have maintained that some very simple concepts closely related to nature are inherited which constitute the brain's basic substrate for the eidetic system when it first confronts its environment (including its own body). The most fundamental, primary, and perhaps primitive inherited concept is the *Ich*, the ego, the self —the basis on which mind constructs its image of the world, as seen through its own means of communication (the senses). This primitive concept and perhaps others must have a strong attractive power on the first raw representations stemming from the physical environment. It is matter-of-fact, everyday experience of the nature surrounding us that needs to be integrated into a logical, empirically ordered set of relationships; for this the self provides the first reference frame (for each individual the center of his universe).

For the transitions between brain space-time and the phase space to occur, however, it is necessary to account for the existence of a logic different from that of nature. Any time sequence of events typical of natural phenomena would not likely represent also the intrinsic logic-based linkage of all the ideas and concepts to which the events belong, or which forms their logical horizon; the latter logic is a property of phase space alone. A natural event can represent only the temporally and causally logical order, and depends strongly on contingent situations —the actual time-bound conditions. Temporal logic and causality constitute a single domain of operation and are concerned with specific time-dependent situations. Therefore, a logic transformation and a corresponding logic operator is needed for the transition from brain space-time to abstract phase space; an analogous, if not similarly structured, transformation and operator perform the transition in the reverse direction, phase space to space-time. I think the importance of the two different types of logic cannot be overemphsized. In conscience phase space the intrinsic logical order of concept-ideas is the logic characteristic of man, the logic which guides his mind. This conclusion leads us once more to consider man as the only being not subject to the laws of nature maintaining its equilibrium. Rather, man is the being that lives in a reversed relationship to nature, not obeying its laws of natural equilibrium but imposing upon it his own will and rationality. However, to express this free will man must discover and then rely on the logical order of the essential and immutable relationships linking the essences, the fundamental concepts and ideas, a logic found only in conscience phase space —not the logic of natural events— however, not directly explorable by mind because it is well beyond the reach of awareness.

As far as consciousness-awareness is concerned, we may envisage two procedures. In the first, consciousness-awareness is stimulated by external events to direct attention to particular lines of thought. This operation may also include a voluntary eidetic component although the need for a decision as to action originates from without. (Voluntary control over eidetic procedures is always in act; at any moment mind can refuse to being directed to or being guided through undesired predetermined thought paths and can divert attention in other directions and toward other goals.) What characterizes this way of working is that, provided a voluntary decision does not interrupt the line of thought stemming from the environment, the same line is causally guided stepwise by the same situation (including updating the development through feedback interactions between the eidetic system and the environment). Whenever the eidetic procedure is concerned only with natural objects, mind-originated lines of development are always bound strictly to the objects' state and the situation, the time development, as it is involved in a continuous interaction with nature. This mode of involving eidetic operations in external events (or objects) forms the core of analytical reasoning. We have characterized analytical thought as continuous interaction between the eidetic system and its objects, eventually resulting in the selection of the best one suited for the situation on the basis of comparisons among possible alternatives. Upon repetitive operations of comparison choices are made along the chain of causal relationships.

In the second procedure conscience phase space presents consciousness awareness with something of a blueprint of the thought to be developed. A major difference here lies in what has been said so far about conscience, the abstract structure comprising all concept-ideas together with their logical, intrinsic reciprocal relationships. The set of essential relationships among concept-ideas of a region of knowledge constitutes the basis of all the lines of reasoning that can be constructed within that region. The logical order among the relationships provides that it can be divided ideally in a number of blocks, each providing all the elements to many possible lines of reasoning. These have a common origin and converge in the ultimate essence of that region, each one representing but one of the manifold aspects of that essential knowledge. (These conclusions, in themselves, do not constitute new assumptions but are contained in the concept of logical order of the essences in abstract phase space and are therefore a necessary property of it.) This type of organization in abstract phase space does not imply any kind of motion here because the logical ordering of ideas can only be as immanent as are the ideas themselves, not subject to change with time since the logic itself which originated it is time independent.

This type of organization in the phase space means that the relationship of consciousness awareness to conscience is the same as that to environment, i.e., consciousness awareness is always "told" by some other entity (whether physical or abstract) what line of thought is to be developed. However, if will is an act related to consciousness-awareness, it must retain its right of vetoing an undesired line of thought and of selecting another line among those present in conscience, in this way asserting its freedom also from the latter. Viewing will as a property of consciousness-awareness alone, we must no longer regard the ideas in conscience as organized in lines of thought, for we have accepted from the beginning that consciousness-awareness can voluntarily develop its own lines of thought on the basis of ideas present in phase space and not necessarily organized beforehand. The objection is not a minor one since it reintroduces the question of the necessity of abstract phase space; here we can advance the same arguments as before. If this organization of conscience depended naturally and exclusively on the fact that all ideas are originally experienced in brain space-time (in the eidetic circuits) according to its temporal and causal logical sequence, conscience would represent merely a duplicate in phase space of what already existed in space-time, and no one would be able to understand the purpose of a conscience at all.

Furthermore, even accepting the existence of a conscience phase space but with component concept-ideas not organically ordered and thus with no clearly identifiable logical relationships among them, will alone would be utterly helpless in selecting the correct items for developing the line of thought in a logically ordered sequence of statements since there would be no logical plan to which refer. If, on the contrary, the phase space comprises not only concept-ideas but also their logical reciprocal relationships (i.e., both the knowledge already experienced in brain circuits and the knowledge in potency not yet realized in space-time and ordered according to the local logic), this phase space is needed since it represents the only source of knowledge not directly linked to previous experience based on the temporal and causal logic of space-time. This phase space is the only source of those hidden relationships for use in consciousness-awareness.

Let us accept that conscience contains all possible knowledge that can be constructed upon the concept-ideas constituting it and their relationships, and that not all possible knowledge can be readily understood by mind (since a considerable part is hidden in relationships among ideas and not retrievable without the corresponding code names). The lines of thought voluntarily selected in consciousness awareness can then be only those that mind is able to select, and these may be very restricted. Thus although free, will may be strongly limited in a way that cannot be predicted because its rationality (in eidetic terms will and rationality are coincident) rests on an incomplete basis and is therefore not truly independent in its choices. This issue continues to draw extensive attention on the part of philosophers and theologians, but neuroscientists have so far contributed little or nothing to the discussion, principally due to their discomforture at pursuing metaphysical reasoning over a topic which they consider appropriate only within the context of positivistic empirical science.

10.5.3 The Hypothesis of the Hybrid Phase Space

The major problem in our view of conscience is with its exclusively abstract character, for this would seem to rule out its having simultaneously a real existence. If idea phase space can be seen as being organized in matter, still enjoying an invariance and a time independence in the sense of immutable objects constantly present (dynamically assured as in a steady state), this phase space would have the character of a hybrid (invariance and abstraction have already been taken as representing a common property). This can be seen in the sense of a space of objects not fixed in matter in a stable way and yet not altogether free from matter. Abstractness here wouldbe "restricted, a property of only one side of the objects, that which the objects exhibit in the interval when they are not organized in matter, and which disappears during the objects' fleeting life in the eidetic circuits. This hybrid phase space certainly would also be a likely candidate for memory, as a continually self-renewing wealth of ideas. Several questions arise from this view of phase space: a) the question needs to be considered as to whether within its matter-bound nature still there is allowed the presence of concepts and systems of relationships not yet understandable by mind although inherent in the concept-ideas; b) the "memory" properties need further examination; c) the relationships with conscience and consciousness need to be reconsidered; d) the relationships between internal and external languages must be carefully considered.

These questions are more subtle than those raised by the notion of a phase space bound only to space and time. Here the phase space would be completely entrusted to the circuits of neurons, where the ideas would be once and for all cristallized in the form of bonds linking the neurons in special circuits, with no other relationships being possible except by creating a new system of material connections. Thus, the physical phase space would represent a rather stable situation, a time independence resting on matter's immutability (though rather precariously), not on abstractness. However, if time independence is possible such as in the hybrid phase space, a degree of abstractness must inhere in the nature of the system. However, abstraction does not reside in the anatomical connections but possibly only at the level of the principle of the system's organization, which allows the appearance of the two aspects of the object, its (reduced) abstractness and its physicality.

Rather paradoxically at first sight, we would find that the system, despite its fundamentally invariant, matter-bound characteristics, can undergo subsequent transformations. In other words, the abstract nature of the missing links among ideas (the concepts and the relationships that mind temporarily cannot understand) may be inherent in a constantly "indefinite" system of bonds between ideas, but one capable of dynamic transformation. Sets of bonds never definitely entrusted to matter in fixed patterns (anatomical and chemical), substitute for the abstract concepts and relationships, but not being physically fixed in a stable, organized structure, they cannot connect the already present ideas in any logically understandable way, and for this reason mind cannot understand all the relationships amomg ideas or form ideas resting on the concepts. For a definitive fixing to occur, a precipitating agent must be available, possibly the missing code name or, more probably, the abstraction procedure applied to the presently existing ideas. (Note that sets of bonds between neurons not physically fixed in a stably organized structure is the only rational basis for developing neural network models of brain matter structure-function.) The conclusion can only be that if matter-bound phase space cannot comprise in one way or another the concepts and all the relationships among ideas, then it may represent the memory. Otherwise, it is much more than memory, the source of present and future knowledge, and cannot be memory (if it were memory, we would find ourselves in the preposterous situation of recalling what our mind cannot understand).

Arguments of a different order must be raised when considering the relationships between the matter-bound phase space and consciousness. If the system of nonunderstandable relationships among ideas is excluded from the matter-bound phase space, the phase space is nonconscience (as previously concluded), and, being in matter, it would at most be related to a state of consciousness-awareness, with more restricted and less embracing properties than simple consciousness when the latter is associated with conscience. It could still be considered as a memory, and would be more strictly concerned with the general mechanisms controlling the level of alertness of the brain (a problem dealt with neither here nor in *PTN*). On the other hand, in a hybrid phase space the system of indefinite connections, presumably related to the concepts and the not yet recognized logical relationships among ideas, i.e., the abstract side of the system, would have no weight in the mechanisms determining the level of alertness. Therefore, a physical phase space, considered as such, i.e., uniquely as phase space, can be conceived only as present in a state closely resembling an "unattentive" stage of alertness, that is, a restricted awareness, but not simple consciousness as we have described it (see Sect. 10.1). On the one hand, sleep can occur when also the restricted awareness (or consciousness) is absent and, on the other, awareness occurs whenever attention is fully aroused and directed to a particular region.

Also the transformation into fixed memory of the hybrid phase space is possible only with the items that mind can understand and use, i.e., simply as ideas. The rest is automatically left aside without the need of filtering procedures by the mind. The brain matter-bound part of phase space is ordered in the form of blocks of ideas according their intrinsic logical (temporal and causal) coherence and coherent with the ideas already implemented in the course of previous and present reasoning. The same hypothesis was advanced above to account for the transition between conscience and consciousness-awareness on the basis of a logical attraction between the contents of these partitions. The significance of the heuristic procedure, obviously not of the actual neural mechanisms involved, can be understood better if we substitute "correlation" for "coherence." Correlation here can be taken in the sense of a measure of the cosine of the angle formed by two vectors, i.e., a measure of their colinearity, which in eidetic operations depends on their reciprocal logical attraction. The logical and the mathematical definitions of correlation thus coincide, and this coincidence helps to provide insight into the logical structure of purely abstract phase space (conscience).

In the hybrid phase space not only ideas but also concepts must be present. For this purpose we said that matter-bound phase space must include an abstract component, which alone can comprise all the objects not yet understandable by mind. If the realization of concepts, together with the ideas which clothe them, from the abstract part of phase space into space and time were merely a question of the code name, memory could still be in the time independent, matter-bound hybrid phase space since its items would be recalled by the appropriate word symbols. These would have to be discovered one by one and could be allowed to vanish when no longer used —without having permanently to fix the configuration of anatomical and chemical bonds that brought the concepts to life— only to be again reassembled when needed. This emphasizes again the need for a phase space at least partly abstract, as the procedures leading to the discovery of new knowledge independent of the surrounding nature would otherwise hardly be conceivable. Only the the existence of an abstract, time-independent component together with a matter-bound one permits such a hybrid phase space all its functional properties as the source of every possible development of thought and of knowledge. Whether a hybrid rather than a purely abstract phase space is more palatable is very dubious. It is just as difficult to envisage the full richness of contents of the myriad concepts in a large number of indefinite anatomical and chemical bonds, the existence of which furthermore remains hypothetical. No less uncertain are the corollary hypotheses of precipitating agents, code names evoking a sudden closure of the anatomical and chemical bonds in the correct circuits. Finally, the arguments in the next two sections concerning language cannot be reconciled with the notion of a hybrid phase space (see next).

10.6 Language and the Eidetic System

Up to this point the problems concerning eidetic procedures have been treated without explicit reference to language. It is obvious that there could be no eidetic activity if the brain did not possess a language, the main characteristic differentiating man from all other living beings. We have suggested above that without an articulated language animal eidetic operations cannot proceed beyond producing rough sensory-based pictures of nature. These pictures represent the milieu in which the animal lives and guide its behavior in its immediate environment and for its life-preserving needs.

In considering eidetic language, one must disregard spoken language which merely translates an eidetic, "internal" language into brain-understandable codes —what we may refer to as "external" language. It is very unlikely that external and internal languages are made up of the same symbols and follow the same logic. Ruling out the possibility of two independent processes, we must see the mind as possessing a means to translate the internal into the external language, a kind of isomorphic transposition from an abstract, unknown form into a material and clearly defined one, and vice versa. A cornerstone of the theory in *PTN* is that the multiplicity of the brain processes are based essentially upon different sorts of transformations, obviously including eidetic procedures and language .

Eidetic activity and language (in its broader sense) cannot be separated basically, as constituting separate functional unities; if we have considered them separately, this was only because language was tacitly taken as an obvious presence assisting eidetic activity, while eidetic activity with all its properties will be assumed from now on as an organizer at the side of language. However, there is no question but that the eidetic activity and language are two facets of a single function since we cannot conceive of the two as existing alone. One generally thinks of language as based on symbols since one uses symbols for expressing words. The symbols constituting external language may nevertheless have very little to do with the internal representation of the same ideas. In thinking one speaks silently with oneself. The necessary transformation from silent to expressed thought certainly occurs at the level of brain cell activity and merely transposes into the motor system —for expressing thought— a set of nerve impulses already present in the eidetic circuits because thought was using the same terms used in brain space-time (that is, as already materialized in the eidetic circuits). This may involve an extremely complicated pattern of transpositions, possibly also enormously difficult to unravel; however, it is conceptually easy to understand in its principle because the organization of the two sets of nerve events is ruled by a same logic, one being the isomorphic transposition of the other. Thinking, in turn, is the result of an operation through which the arbitrary code names indicating the ideas become linked to one another by other code names, the latter being the logical ordering elements of every thought, without which we could not make us understood even to ourselves. Only through the ordering symbols does the whole reasoning take a logically analyzable form, so that thought, through analysis, may obtain its necessary self-explanation. The whole procedure can take place only in the stage of awareness, as function of space-time. Accordingly, once more we must ask what relationship the language, so organized, has with the immanent world of conscience, where there must also be a logical order of the concept-ideas in lines of thought, here according to an internal language about which we can only speculate.

External language, as it can be empirically objectivated, is thus a property of space-time, i.e., of the cerebrum, because its being conceivable and understandable makes it an object that can exist only in space-time. On the other hand, that which constitutes the language in the phase space of immanent conscience cannot be the object of direct knowledge. In brain space-time we know only the result of the transposition of what we assume to represent a fundamental, abstract form of language, integral to phase space, the necessary existence of which we have already accepted. With the acceptance of its existence, however, we must also accept that in the transposition into consciousness-awareness it becomes clothed in the same symbols used for brain-internal and interindividual communication. Let us consider a few arguments that may help us to understand the nature of the phase space internal language. As far as we know, learning the external language depends largely on the sensory inputs to the brain; we need think only of the difficulties in learning all types of human communication in cases of congenital or early acquired deafness. Every sensory deficit impairs learning the association between symbols and the corresponding objects (physical and abstract) in consciousness-awareness before being translated into conscience. This in turn hampers the development of the eidetic activity that can begin only in space-time. In fact, since everything coming from without must strike the sense organs, all sensory inputs must first be translated into neural symbols (coded nerve impulses), the operation making possible their subsequent use in the brain.

The relationship between natural events and sensations is analogous in a way to that between internal and external languages. In nature, for instance, the electromagnetic radiation within a range of frequencies induces changes in pigment molecules of the eye's retina and consequently in the connected nerve cell structures and gives rise there to nerve impulses, until at the end of a chain of neural events we eventually come to "see" lights and colors. However, these are abstractions and as such are completely foreign to nature. In this case, too, we seem to have a transposition of languages, in the sense of translation of information, but the natural event —the electromagnetic radiation— does not communicate through words or any other symbolic language. Our sense organs provide the necessary transformation of the characteristics of the natural events, such as the energy-frequency of radiations or other different properties of nature, into another class of events which our brain is able to understand, the nerve impulses representing the basis of brain language, of external language. Something of this sort occurs in the language transposition between consciousness and conscience. The nerve impulse coded sequences making up the language of brain are transformed in the transition to conscience into an immanent "unknown," part of conscience phase space but not existing as such outside phase space, as the light does not exist outside the brain. The light that we see is the cerebral interpretation of a natural event made up of electromagnetic radiations at some frequencies, the latter being the true, essential nature of light. Likewise, we may say that the immanent unknown, which is part of conscience phase space, is translated in the transition to consciousness-awareness into external language in the form of an idea's code name. We may say that external language is the cerebral interpretation provided by a transformation in consciousness-awareness of that which is actually a constituent of conscience. However, while we know the essence of light, the electromagnetic radiation, what *is* the phase space conscience? What is the essential nature of conscience? What is the essence of the abstract structures constituting it, their logic, their order of relationships?

Lacking a better term for it, we shall continue to speak of an internal language to denote the result of this transposition or transformation into conscience phase space and having the properties of an essential language (or a language of essences). It is commonly held that the formation of ideas is facilitated by language, but I would maintain that language is the sole basis of eidetic activity. Certainly, however, through an internal language, as articulated and precise as possible (if in conscience there is an essence language, this would by definition be precise and all-comprising), we cannot say that the formation of ideas is simply speeded up. That which we call here the true nature, or essence, of internal language can only be the true nature of what the symbols of external language are intended to convey, i.e., the concept-ideas themselves, which may become objects of our thought only when clothed in these symbols. We have maintained that space-time eidetic activity, to be able to construct ideas (those not directly concerning natural events) must first rely on the concept as the non-symbolic source of the ideas. The concept is the sole abstract object which is primarily "given," or which imposes itself on understanding as a self-sufficient and self-explanatory reality, to be transposed into the symbolic language of communication. The concept must be clothed in ideas which introduce the concept into space-time, since these form the essential basis of external language.

If we did not have such an internal, purely abstract language at our disposal, we could probably form only rough substitutes for the concepts, akin to blurred sensory images, or painful representations, etc. We would not be able to abandon the brain-external language simply made up of nerve impulses —not symbolic ideas— and could therefore not construct true abstractions, i.e., not build a real phase space. These rough substitute representations might also be naturally associated to one another in some way, without producing, however, an autonomous reasoning, only sequences of such sets of impulses substituting for the ideas, ordered only according to the temporal and causal logic of the relationships among the physical objects of the immediately surrounding nature. In this way, a cat can also have a "knowledge" (obviously not a conscience), for instance, of the house as the location for shelter, food, defense, and care and acquire an "awareness" of the inhabitants as friends, and of all others as foes. Therefore, the real difference between humans and animals is man's ability to transform nerve-coded signals into the symbols of highly articulated external language and then into the essence of that which the symbols represent, i.e., into abstract structures, the ideas.

However, since the ideas exist as such only in the form of abstract objects, for them to exist also in our awareness they must have a definite shape in space-time, i.e., they must be embodied in brain circuits, the only form that allows us to use them and through them to obtain a glimpse of their true origin, the higher abstractions, the concepts which comprise the ideas. These features are altogether absent in animals. What animals can have at their disposal are simply the first, nonmediated, and primitive representations of external or inner situations as they are transduced by sense organs into nerve impulses. We might call these the "machine language" —to use computer terminology— not originating in the transition from phase space to consciousness-awareness but at the first, lower stages of brain operation, strictly related to nature's materiality, to the visible and tangible in the surroundings, to what talks to the brain through the senses. The capability of man's eidetic system for two-way translation between an external and an internal language provides him with the means for developing the most complex abstractions, such as those of metaphysics and of mathematics; by exploring his own thought, man has pursued philosophy, science, ethics, and religion. (These and other related points are developed in the "Appendix" of *PTN*, in connection with the differences between man's and animal's intellectual powers, where a number of neurophysiological and metaphysical conclusions are drawn.)

10.6.1 Language Transformation in Realizing Conscience

We have outlined conscience's properties in considering it as a phase space in which internal language and ideas become fused and represent the unity of nonsymbolic and symbolizable aspects of a conceptual essence. However, this space does not contain the symbols of space-time used in consciousness-awareness because that which constitutes phase space conscience —concept-ideas and internal language— are purely abstract objects that cannot be divided in terms of time or space. This means that the translation into consciousness —to provide all the symbols in which the abstractions of the phase space ought to be clothed, i.e., the ideas to be expressed in the external language of consciousness-awareness— is a true transformation providing a space-time reference for these abstractions, which by themselves do not know physical space and time. Performing such a feat becomes possible only if something having the nature of a "description" (resembling an identification card) of those essences can be found, and this can only be the symbolizable aspect together with the nonsymbolizable aspect of the same essence.

Nothing can be said concerning the nature of this description or that of its symbols (if ever there is any). However, for the sake of comprehensibility, let us return to our previous analogies. For instance, the essence of light, radiation, has a description in its frequency spectrum, as does odor in its molecular composition. Likewise, we may think of every autonomous and self-explanatory reality, i.e., every concept of phase space, as being partly describable through the essence-elements, that is, the ideas comprised by the unitary essence. Only an essence not comprising other essence-elements, an indivisible, noncompound unity, would not be describable in any way, that is, it could not be precisely named and described in our space-time by external language; this unity would remain forever closed in conscience, never graspable in any way by the brain, and capable of present itself only as "present." (The metaphysical significance of this conclusion differs according to philosophies and beliefs.) This is important in the context of the discussion concerning "essence" (the *essentia*) and being (the *esse*). The *esse* is what makes a thing *be*, i.e., a pure "act" without attributes; the essence is what makes a thing be *what it is*. Since the "what it is" of each thing is defined by its properties or attributes, differing from those of other things, the properties and attributes are the description constituting the essence. However, this must always be referred or related to an *esse*, an individual one, to an act, i.e., a reality, the real core without which no essence could exist, as no attribute or property could really exist alone. The real existence (as "act") of the indivisible, noncompound unity is that which actually makes real the conscience phase space as comprising all the essences since these really exist as pertaining to really existing *esse*. Between the pure act (the pure *esse*) and the essences pertaining to it in conscience phase space, we can imagine a transition term, a bridge having, on the one hand, the characteristic of not being describable (as the *esse*), at least in part, and, on the other hand, of originating incomplete descriptions of itself through a manifold of ideas. This transition term is the concept.

The necessity of a transformation between matterless, timeless conscience and space-time consciousness-awareness is again emphasized by what we may experience as suddenness of appearance of ideas in the space-time of eidetic circuits; the idea appears to us as an "immediate" synthesis of an essence from a set of element components. However, we should not let ourselves be misled by the immediacy represented by the symbolization of the idea in a simple, single word, that is, we must not confuse immediacy with simplicity. The name for the idea does not come out of the (abstract) blue because at the time that the word appears the transformation between phase space and space-time is already completed. The same possibility of using a word is the demonstration that we already are in space-time and we use spatio-temporal forms of thought, i.e., words. External language is sometimes admirably concise, depending on the way in which it is conceived, but it still betrays its artificial, conventional nature as a means for indicating (naming) an essence, and only as a space-time procedure. As we have said, in phase space internal language is the same thing as concept-ideas and their descriptions; these, however, do not give themselves away in their entirety in the transformation into spatiotemporal events, so that the whole, unitary, real content of a concept-idea could never be completely contained in a single word. This implies that the two descriptions, in phase space and in space-time, must be profoundly different in nature.

The temporal aspects of the question become particularly clear considering the way in which a new constructed idea is assembled from concept-ideas belonging to different regions of the eidetic system. Here the spatiotemporal procedures developing in consciousness-awareness by themselves would have no way of symbolizing beforehand the constructed idea since they could not know what the idea ought to look like before constructing it. Moreover, to be able more or less precisely to name it, since this presupposes a definition, the spatiotemporal procedures ought to be able to grasp the idea in its entirety, i.e., together with the description, which is possible in phase space only. Therefore, only through the suddenness of a suitable transformation from conscience to consciousness-awareness does one obtain the flashing translation of the symbolizable part into a space-time representation of the idea as embodied in eidetic circuits; for this worded symbols may be found that are as concise as possible and, although having some meaning, remaining no more than a ghost of that concept hidden in phase space.

One might be tempted to assume the velocity of internal "vision" of the ideas translated from conscience phase space into consciousness space-time as the comparison term for the velocity of the events developing in brain space-time. However, one must be careful not to misjudge the situation. The transformation from conscience phase space, making the concept-ideas appear (though not in their entirety) in consciousness space-time, can only be of a very special kind indeed. According to their definitions, the first space is altogether abstract, a nonmatter and nontime space; the second is a physical space of matter and time. The transition from the first to the second can be said to be operative only at the very moment that the ideas appear in space-time. Actually, the transition itself does not exist up to the moment when the abstract structures become part of space-time. Until that moment there is no time, only timeless abstractness and therefore no procedure of transformation; from that moment on, the abstract structures are no longer abstract but are "living" a material life in the brain. This means that only from this moment on we speak of time. Thus, the transformation is not a time-consuming procedure but a timeless (or, suddenly instantaneous?) change of the "quality of reality" enjoyed by the concept-ideas, from the merely abstract to the actually physical state. However, the abstract object, the concept-idea at the origin is maintained unaltered, although imprinting some of its form onto the structure embodied in matter, endowed with the physical, if fleeting, reality.

We have said that the two states of the concept-ideas —abstract and physical— coexist since the first is timeless and immanent, while the other comes to a fleeting existence only when needed, without altering the former's state. The two may thus be considered as belonging to a single system, itself having the possibility of being simultaneously in both states, not according to a mere probabilistic principle (as in quantum mechanics systems) but because the eidetic system at every moment is prodded to communicate with itself, with the rest of brain, and with the outside world. This does not differ from considering the physical state as being imperfectly in act what the abstract state is perfectly in potency. In both pictures, however, the phase space conscience and the space-time consciousness are much more closely linked than it may have appeared until now, and from this point of view the conscience phase space possesses a not easily questionable reality. Now, since the suddenness of appearance of the ideas in the state of consciousness-awareness cannot be taken as a reference for actual brain space-time processes because no change in space and time can be compared with a nontime change, the transformation from conscience phase space to the space-time of neuronal circuits is clearly a true transformation of reference framework. As stated above, the transformation of reference framework cannot take place from phase space to awareness directly but is mediated by simple consciousness, which acts as the two-way "door" between conscience and awareness. Further transformation from simple consciousness to awareness, by simultaneously activating all the feedback circuits needed to construct an idea, is the one assuring the quasisuddenness —the high velocity— of appearance in awareness. But, here the suddenness is only in relation to the other, much slower brain processes since consciousness is already a function of physical space and time, not logically comparable to abstract space and nontime.

10.6.2 The Nature of Language

To understand the role of phase space —the region of immanent concept-ideas— in the eidetic system one must regard the phase space as possessing language. Pushing this a step further, we can say that conscience is itself *the* language (the only possible, internal one), whereas consciousness-awareness possesses merely *a* language from among the many possible ones. Actually, the latter can be very differentiated, as differentiated as are the languages of different peoples. However, it would be difficult to maintain that concepts (and related ideas) in their essences differ substantially in the different individuals or peoples. Consequently, the ability rationally to connect concepts and ideas, as they are in phase space, according to all their possible logical bonds, cannot be property of any one particular language; rather, it must be the peculiarity of a unique common language to all peoples. It must belong to the essential nature of conscience, the "language of logic," or logic per se; in this way it is the outstanding characteristic unique to all humans, and humans alone. This conclusion is reinforced when we consider that the eidetic system, particularly in the case of constructed ideas, utilizes concepts and assembles elements and ideas in different regions of knowledge. This in turn, is integral to conscience, for it results from transformation into phase space, after being shaped as ideas, of the preceding space-time neural operations performed on brain "experiences" (themselves transformations of what the external world presents to the brain senses) and integrated with its concept-ideas. Phase space conscience and knowledge, conscience and internal language, and therefore knowledge and internal language, coincide, each being but one aspect of the manifold abstract essence of mind.

The transformation from phase space to space-time, besides providing a time reference for the transposed items, also provides the new logical reference. We have said that the order of the items (concept-ideas) constituting the phase space knows neither time nor matter. That is, the logic of that order must be different from the logic in space-time, which is the order of the objects with a past, present, and future as well as a location, in changeable causal relationships to each other in space and time. The objects constituting the phase space, on the other hand, cannot possibly be subject to a "here now" and a "there tomorrow" and can obey only the logic of an invariant abstract structure, remaining immanent as essences and as relationships between the essences. Thus, the transformation must provide the logical key by which an order may be transposed into a different one in essence, as different as the essence of the two spaces and their constituting objects. The logic which defines the relationships among the objects of space-time and their constituting elements forms the basis for their description in the causality domain and is the logic of language in space-time terms, the external language. Therefore, since this logic may be appraised differently by the individual minds due to a variety of outer conditions and inherited properties, external languages —logic in space-time— differ to varying degrees among large classes of individuals according to geography and race. This kind of variability cannot be found in the logical order of phase space because its logic is the logic of the immutable relationships between essences, not subject to different appraisals by different minds since the aware mind, that part of mind living in space-time, cannot directly explore and thus cannot directly gain knowledge of the objects of phase space but only of their imperfect transposition in space-time, under another logical order.

Conscience phase space is the region of internal language —one might say of the "essential language" or, even better, of the "matrix" of all languages— and is the origin to all external languages of thought in consciousness-awareness. If this were not the case, we would not be able to understand why only humans have language and consequently rational thought —language and rationality being derived by a transformation from phase space logic. Man was the last mammal to appear, but the brains of all mammals —since the first of these during the Paleocene some 70 million years ago— are generally comparable in terms of overall structure to the brains of present mammals. And from the Pliocene, about 10 million years ago, when "hominization" began until today, the nervous systems of other mammals did not develop further, or, if a progressive development did occur, this was at a much lower pace than in humans. As repeatedly stated, the brain of monkeys (to say nothing of cats and dogs) is unable to create constructed ideas as man's brain does and is also unable to group and order ideas in an autonomous, logical reasoning; this remains man's peculiarity. Yet, the overall size of primate's brain and the complexity of its feedback circuits may not differ substantially, as far as we can tell, from the human's, in which ideas are implemented.

If one wishes to describe as eidetic only the consciousness-awareness function, that is, physical space-time, one cannot understand why eidetic language does not manifest itself in monkeys. The only signs that we obtain from the latter are a handful of crude vocalizations and gestures, the very primitive and lower grade manifestations of fear, rage, pleasure, pleading, and such. The presently insufficient comparative knowledge of the fine structure of brain circuits prevents us from pushing these arguments further, but the principal facts appear sufficiently evident. Obviously, the question cannot be settled here. I simply wish to point out the multiplicity of arguments leading us to consider as necessary the existence of a region in which the eidetic function originates simultaneously in all its many facets —phase space of ideas, knowledge, logic, and language matrix. This also applies to will (accepting the arguments put forward in the preceding sections) although it is a function of consciousness-awareness in terms of its manifestations. Its origin, i.e., its logic (and the dependent rationality), has its deepest root in conscience phase space and receives its primary justification from the immanent, immutable, essential logic, the internal language. In the transformation from phase space to space-time the *Ich*-will assumes the space-time autonomy of thought, the freedom of eidetic procedures guiding behavior (see Sect. 10.3.1). However powerful the logic in phase space concepts may be, it can never be precisely and completely transposed into brain-external language; therefore man is left to confront everything which his brain is able to understand of the still obscure concepts of conscience and of the "primary motor" *Ich*, with his own judgement and decisions elaborated in consciousness-awareness. Therefore, man's free will, which can oppose the logic, is the offspring of space-time brain.

While gross fundamental structures do not distinguish the different mammalian brains, what does materially distinguish them is mainly the amount of nerve cells, suggesting a much greater complexity of interconnections reached with evolution, whereas the fundamental principles of the "building" remain similar. As stressed in *PTN*, the basic element of differentiation is, more precisely, the complexity reached in organization, making use of units consisting of basically rather similar architectonic networks but from time to time assembled according to different patterns. Certainly, in order to display its present enormous possibilities, the organization of the human brain has also required an increased richness of structural details, but this does not consist essentially in the architecture or overall pattern of the system (as a skyscraper with a thousand apartments does not differ in "essence" from a common two-story house). Organization means differentiation of capabilities based on the use of the same elements in differently articulated networks, according to purposes. The fundamental and most striking of these is language, made up of logically connected ideas and sentences that we propose as being organized in matter under the control of a logical, abstract matrix, part and parcel of conscience phase space.

The language transformation occurring between phase space and space-time acquires a primary, outstanding place in the history of evolution. It is suggested in *PTN* that language is the main factor behind maturation of the human brain. We can now say that the appearance of the language transformation may have represented the critical factor in the differentiation of brain organization and promoting the development of the eidetic system which characterizes humans. Only through external language can concept-ideas be "focused" upon, analyzed in brain space-time, i.e., rendered in a more or less precise and articulated form, and ordered according to their logical space-time relationships. However, as a rational power —say, as expression of logic— this capacity must preexist to, or at least must exist independently of, the actual content of the eidetic constructions in space-time, as the exclusively human capability for giving a description to the essences of phase space and to the relationships linking them to one another. Since the description is not intended to consist of the actual word with which one signifies the object in space-time but what we have defined as the symbolizable, descriptive part of every essence, i.e., of every concept-idea and of its relationships, the capacity of logically linking concept-ideas can be a property only of the phase space and is alone able to make conscience the region of ideas, of integrated, simultaneous knowledge (as "conscience" actually means). For this reason, conscience must have a structure of its own, what we have called the "logical matrix" of internal language which, through the language transformation, eventually also induces organization of brain matter, i.e., of space-time elements from which the logic of time and causality of external language arises.

Our reasoning has become repetitious. By retracing the path of eidetic activity from reasoning to the formation of the ideas, from the latter to concepts, to language, and to logic matrix, we have reached the source of the eidetic function itself, the origin of the logic with which our mind operates. Unfortunately, we cannot specify any neural mechanisms involved in this, or give a name to the neural processes underlying the eidetic procedures, besides those of comparison and, possibly, what I have called the implicit implementation of ideas in feedback circuits. Experiments, with today's methods of neurosciences, can only produce evidence of neuronal involvement in these procedures; but the simplest of ideas, as such, still escapes any experimental, material test. Certainly, we must yet define a better set of coordinated operational hypotheses to test a theory of the neurological principles at the basis of the function.

I hope that the arguments put forward here may prove sufficiently convincing to entice other investigators to develop more sophisticated approaches than those now available. With the present state of knowledge, our efforts would have been doomed to sure disappointment and failure had we been looking with today's physical and chemical methods for the neural mechanism through which a language transformation between conscience phase space and brain matter space-time transposes the organizing power of language from the region of abstractness (the logical matrix in phase space) into the external language of physical reality. The point that I have tried to make, however, does not concern the practical approach to these problems. The core of the question remains the resistance to accepting a conscience phase space as a real entity: matrix of language and logic (if the two can ever be conceived as separate entities, or properties) are integral to the latter inasmuch as they represent its abstract substance. Obviously we could make the matter somewhat easier by saying that many (certainly not all) of the properties that we attribute here to conscience phase space are those commonly attributed to the so-called "subconscious" self —an obscure entity, whose existence, however, no one seems to doubt and believed to relate exclusively to space-time. In this way, phase space would still be anchored in space-time and would manifest itself exclusively in neural activity that we can always manage to monitor. Even so, the so-called "subconscious" would unfortunately represent but a small fragment of the overall problem, the part that in one way or the other can be "seen," leaving shrouded in darkness that which produces and organizes it. The "form" of that organization must necessarily "exist" and have a "location." I am thoroughly convinced that we cannot escape, by any sort of twisted reasoning, the confrontation with the "reality of the abstractness" inherent in eidetic activity. As the mathematician has helped the physicist to foresee and track down subatomic particles and the astronomer the invisible celestial bodies, I am convinced that the true aim of our endeavor is to first find the substitute in the analogy for mathematics to be used with eidetic processes, i.e., the fundamental logic and language with which we can understand —say, mentally explore— the essential abstractness of mind, the conscience phase space.

References

The following references, besides the few quoted in the text, concern only those readings that gave the author particularly much to think about, selected from a great number of articles and books in the decades from the late 1940s until now, which have been searched for new ideas, for hints and suggestions, for openings in still unexplored regions, and constituting the cultural background of this book.

Neurophysiology and Related Subjects

Adrian ED (1947) The physical background of perception. Clarendon, Oxford

Arduini A (1987) Principles of theoretical neurophysiology. Springer, Berlin Heidelberg New York Tokyo

Basar E (1983) Toward a physical approach to integrative physiology. I. Brain dynamics and physical causality. Am. J. Physiol. 245: R510 R533

Basar E, Flohr H, Haken H, Mandel AJ (eds) (1983) Synergetics of the brain. Springer, Berlin Heidelberg New York Tokyo

Blakemore C, Greenfield S (eds) (1987) Mindwaves. Blackwell, Oxford

Bleuler E (1960) Lehrbuch der Psychiatrie. Springer, Berlin Göttingen Heidelberg

Churchland PM (1984) Matter and consciousness. MIT Press, Cambridge

Churchland PS (1986) Neurophilosophy. MIT Press, Cambridge

Crooks JH (1980) The evolution of human consciousness. Clarendon, Oxford

Edelman GM (1989) The remembered present. Basic Books, New York

Elsasser WM (1958) The physical foundations of biology. Pergamon, London

Fraisse P (1984) Perception and estimation of time. Ann Rev Physiol 35:1-36

Jaspers K (1959) Allgemeine Psychopathologie. Springer, Berlin Göttingen Heidelberg

Klatzky RL (1985) Memory and awareness. Freeman, New York

Pellionisz A (1983) Brain theory: connecting neurobiology to robotics. Tensor analysis: coordinates to describe, understand and engineer functional geometries of intelligent organisms. J. Theor. Neurobiol. 2:185-211

Rakovic D, Koruga Dj, Djakovic D, Martinovic Z, Desimirovic V, Minic Dj (1989) Ultralowfrequency "optical" biocomputers: biophysical arguments. In: Hong F (ed) Molecular electronics: biosensors and biocomputers. Plenum, New York

Rashewsky N (1960) Mathematical biophysics. Physico-mathematical foundations of biology. Dover, New York

Richelle M, Lejeune H (1980) Time in animal behaviour. Pergamon, Oxford
Von Neuman J (1959) The computer and the brain. Yale University Press, New Haven

Physics and Related Subjects

Born M (1964) Die Relativitätstheorie Einsteins. Springer, Berlin Göttingen Heidelberg New York
Brillouin L (1956) Science and information theory. Academic, New York
Denbigh KG (1958) The thermodynamics of the steady state. Methuen, London
Dirac PAM (1981) Principles of quantum mechanics. Clarendon, Oxford
Einstein A (1916) Die Grundlage der allgemeinen Relativitätstheorie. Ann Phys IV F 49:769-822
Einstein A (1969) Über die spezielle und allgemeine Relativitätstheorie. Akademie, Berlin
Gal-Or B (1981) Cosmology, physics and philosophy. Springer, Berlin Heidelberg New York
Heidmann J (1980) Relativistic cosmology. Springer, Berlin Heidelberg New York
Honerkamp J, Römer H (1986) Grundlagen der klassischen theoretischen Physik. Springer, Berlin Heidelberg New York
Katchalsky A, Curran PF (1981) Nonequilibrium thermodynamics in biophysics. Harvard University Press, Cambridge
Khinchin AI (1949) Mathematical foundations of statistical mechanics. Dover, New York
Lichnerowicz A (1967) Éléments de calcul tensoriel. Colin, Paris
Lindsay RB, Margenau H (1957) Foundations of physics. Dover, New York
Prigogine I (1967) Introduction to thermodynamics of irreversible processes. Wiley, New York
Schmutzer E (1981) Relativitätstheorie-aktuell. Deutsch, Frankfurt
Shannon CE, McCarthy J (eds) Automata studies. Princeton University Press, Princeton
Straumann N (1981) Allgemeine Relativitätstheorie und relativistische Astrophysik. Springer, Berlin Heidelberg New York
Wichmann EH (1971) Quantum physics. Berkeley physics course, vol 4. McGraw-Hill, New York
Wiener N (1961) Cybernetics. MIT Press, Cambridge; Wiley, New York

Philosophy and Related Subjects

Bleuler E (1963) Das autistisch-undisziplinierte Denken in der Medizin und seine Überwindung. Springer, Berlin Göttingen Heidelberg
Carnap R (1958) Introduction to symbolic logic. Dover, New York
Chenu M (1963) L'homme dans le monde. In: Association des Professeurs de Philosophie des Facultés catholiques de France (eds) Recherches de philosophie. VI. Saint Thomas d'Aquin aujourd'hui. Brouwer, Paris, pp 171-175
Coggi R (1988) S.Tommaso d'Aquino. Pagine di Filosofia. Edizioni Studio Domenicano, Bologna
Czeschlich D (1987) Irrtümer in der Wissenschaft. Springer, Berlin Heidelberg New York

D'Espagnat B (1983) In search of reality. Springer, Berlin Heidelberg New York

Dubarle D (1963) Cosmologie tomiste et philosophie naturelle contemporaine. In: Association des Professeurs de Philosophie des Facultés catholiques de France (eds) Rechercehes de philosophie. VI. Saint Thomas d'Aquin aujourd'hui. Brouwer, Paris, pp 137-169

Fabro C (1963) Le retour au fondement de l'être. In: Association des Professeurs de Philosophie des Facultés catholiques de France (eds) Rechercehes de philosophie. VI. Saint Thomas d'Aquin aujourd'hui. Brouwer, Paris, pp 177-196

Hegel GWF (1967) Phenomenology of the mind. Harper and Row, New York

Hegel GWF (1969) Philosophy of right. Oxford University Press, New York

Husserl E (1983) Ideas pertaining to a pure phenomenology and to a phenomenological philosophy (1950-52). Nijoff, Den Haag,

Hyppolite J (1974) Genesis and structure of Hegel's phenomenology of spirit (1946). Northwestern University Press, Evanston

Nagel E (1961) The structure of science. Harcourt, Brace, New York 1961)

Petrus de Bergomo (1960) In Opera Sancti Thomae Aquinatis index. Paulinae, Roma

Popper KR, Eccles JC (1977) The self and its brain. Springer, Berlin Heidelberg New York

Thoma de Aquino (1962) Summa Theologiae. Paulinae, Roma

Sertillanges A-D (1925) S.Thomas d'Aquin. Alcan, Paris

Binding: Buchbinderei Lüderitz & Bauer, Berlin

MIX
Papier aus verantwortungsvollen Quellen
Paper from responsible sources
FSC® C105338

If you have any concerns about our products,
you can contact us on
ProductSafety@springernature.com

In case Publisher is established outside the EU,
the EU authorized representative is:
Springer Nature Customer Service Center GmbH
Europaplatz 3, 69115 Heidelberg, Germany

Printed by Libri Plureos GmbH
in Hamburg, Germany